T0186952

The Science of Fate

Dr Hannah Critchlow is Science Outreach Fellow at Magdalene College, University of Cambridge, and has been named a Top 100 UK Scientist by the Science Council for her work in science communication. She is listed as one of the University of Cambridge's 'inspirational and successful women in science' and appears regularly on TV, radio and at festivals to discuss and explore the brain.

The Science of Fate

*The New Science of Who We Are – and
How to Shape our Best Future*

Hannah Critchlow

HODDER

First published in Great Britain in 2019 by Hodder & Stoughton
An Hachette UK company

This paperback edition published in 2020

6

Copyright © Hannah Critchlow 2019

A CIP catalogue record for this title is available from the British Library

Paperback ISBN 978 1 473 65931 5
eBook ISBN 978 1 473 65930 8

Typeset in Bembo by Palimpsest Book Production Limited,
Falkirk, Stirlingshire

Printed and bound in Great Britain by Clays Ltd, Elcograf S.p.A.

Hodder & Stoughton policy is to use papers that are natural,
renewable and recyclable products and made from wood grown
in sustainable forests. The logging and manufacturing processes
are expected to conform to the environmental regulations
of the country of origin.

Hodder & Stoughton Ltd
Carmelite House
50 Victoria Embankment
London EC4Y 0DZ

www.hodder.co.uk

To baby Max, watching your destiny unfold is a wonder

Contents

CHAPTER ONE

Free will or fate?

One stifling day, at the beginning of the long hot summer of 2018, I sat in the waiting room at my GP's surgery. Outside it was dazzlingly bright, but inside the fluorescent lights were still humming. A buoyant doctor strode out and called my name. I took hold of my two-year-old son's hand and we followed her down the corridor into a small room where she took a sample of my blood. The vial contained thousands of white blood cells. Hidden inside each one was my DNA, the 3.2-billion-lettered code unique to every human being that is the blueprint for life.

My son and I were at the hospital because my father had been diagnosed with haemochromatosis, an inherited condition in which iron levels slowly build up in the body. Eventually the excess iron begins to damage internal organs and, if left untreated, it can lead to heart disease, diabetes and cirrhosis of the liver. Thankfully, in my father's case, the organ damage was not that far advanced, but because the condition had gone undiagnosed for decades he now has to undergo weekly bloodlettings. This treatment, while intrusive, means he is otherwise in good health. A happy outcome for him and those of us who love him.

Since the disease is genetic in origin, the NHS offers a test to other family members who may be affected. That means me, my sister, cousins and, potentially, our children. It's a simple blood test and the result comes back quickly. On the face of it my decision should have been straightforward. Neither my son nor I needed to know yet whether we, too, carried the gene variant that causes haemochromatosis but at some point we would have to find out. If we were positive we would need to reduce our consumption of iron-rich food and the levels of iron in our blood would need to be carefully monitored. Testing wasn't urgent but it couldn't be put off for ever.

I am a neuroscientist and have been fascinated by the idea of biological determinism all my working life but I struggled with this decision far more than I thought I would. I practised detachment, reminded myself that I believe knowledge is power and that understanding my own body is the most empowering knowledge of all. But still I didn't book the appointment. I knew that if I tested positive, I would feel compelled to read all the scientific literature to come up with a plan for changing our lifestyle. Would it feel like an extra anxiety-making responsibility or would it empower me to make changes?

I found my opinion changing day by day. In the end the decision was simplified by the GP informing me that the NHS would not consider analysing my son's blood unless I came back with a positive result myself since there was no known history of the condition in his father's family. I went for the test in order to discover the risks for myself, and potentially also my son. But it took me weeks to pick up the result. I was surprised by how challenging I found it to step confidently into a position of knowledge when it concerned something so intimate, especially where I would then have to decide for my child. It felt unsettling as well as alarming.

In the end, the results showed 'a heterozygous genetic variation', which means that I'm a carrier but unlikely to develop symptoms.

I hadn't anticipated this scenario and, though I felt a small measure of relief for myself, I felt frustration that there could be no definitive peace of mind about my son. He, too, might be affected by the condition in the future but, given my result, the test would not be offered to him on the NHS unless, and until, he showed symptoms. The whole episode was a lesson in the emotional nuances of what I had previously thought of as a straightforward practical question, and an enlightening context from which to pursue my investigation into the extent to which any one of us is free to determine our fate. I have been left a bit humbled by my sense of our brush with an implacable force.

Since the dawn of humanity our species has been trying to figure out what, or who, is calling the shots. The question of whether we can determine our life's course or should accept that it is largely beyond our control is well up there on our list of thorny conundrums to resolve. Are we fully conscious agents possessed of free will or closer to pre-programmed machines, running on deep drives of which we may not even be aware? At different times and in different places human beings have answered this question in many ways. We've asserted that we're animated by a divinely bestowed soul, or inspired by the quasi-godlike powers of our own mind, or powered by the neurochemistry zapping round our brains. Whatever the flavour of the answer, the problem of whether or not we can steer our own way arises directly out of being an animal in which consciousness is so well developed that it enables us to ponder consciousness itself.

This book considers the question by applying insights from the discipline of neuroscience. Modern medicine has shown us that what we put into our bodies interacts with our individual genetic inheritance and gives rise to outcomes: low blood pressure for some, high cholesterol for others, haemochromatosis in my father's case. Increasingly the brain is being viewed in the same way. It processes incoming signals via circuits that have been laid down by our genetic inheritance, and those complex processes give rise to outcomes in the form of thoughts, decisions and choices. I wanted to address the possibility that at the intersection of innate features common to all humans, and the genetic package unique to each individual, something like a twenty-first-century version of fate is generated.

As far as many ancient cultures were concerned, fate or destiny was definitely all-powerful. The Ancient Greeks believed that even the gods could not escape it. (Not much wiggle room for us mere mortals, then.) In the era when monotheistic religion dominated, God was the ultimate decider of any individual's outcomes. These days, at least in Western post-industrial societies, the secular majority of us assume that we ourselves are the authors of our own life story. We may still speak of someone being 'destined for greatness' or 'fated to fall in love' but, to our modern minds, fate is really nothing more than a figure of speech. Many of us go about our lives believing that though there are undoubtedly constraints, such as the country, class or race we happen to have been born into, within that context we are free agents. We can make our choices – from what to have for breakfast to selecting our friends and opinions – based on rational decision-making processes; over time these choices evolve into behaviours and habits, and eventually the collection of experiences that constitutes our lives.

We use memory, language and narrative to rationalise our life and shape it into something we understand and feel we can control. It makes perfect sense that we should do this, of course. We live inside our staggeringly sophisticated minds just as much as we live in our bodies, so our own selfhood sits at the centre of the universe we perceive around us.

But even though we operate from day to day as if our conscious mind was the undisputed captain of a quiet ship, we probably all know, deep down, that it isn't quite that simple. The mind is a much wilder place than that, where conscious decision-making is only a fraction of the full story. Human society has always been afraid of unconscious forces, stigmatising them as threatening, even demonic. Anybody who has ever experienced or witnessed mental illness will be aware that the mind can feel very alien or even downright terrifying. But to conceive of the unconscious as a danger zone that must be rigidly policed is to misunderstand the essential role it plays in everyday life. As we will see later, a great deal of decision-making and routine judgement takes place without us being aware of it. If it didn't, we would scarcely be able to function. The time-consuming effort of consciously instructing every decision and evaluating every situation would keep us flailing around, trying to get out of the front door long after we were supposed to be at our desks.

Most of us probably do not believe that we are fully rational beings who can do or choose whatever we like and steer our own outcomes as we will. We accept, more or less readily, that as well as powerful subconscious forces, external factors shape and, to some extent, determine our lives. Fate may have fallen out of fashion but many of us will concede that luck, good or bad, has played a part in our story. We were in the right place at the right time to meet our future spouse or land a dream job. Serendipity allowed us to encounter the friend who could help us resolve a dilemma, or a cruel twist meant we missed an opportunity that we feel sure, in hindsight, would have changed the course of our lives.

And most of us are comfortable with acknowledging the role played by other people and our environment – our family background, our education and early life experiences – in shaping our personality and outcomes. It is commonplace, for example, to say that being brought up in a loving family, or a neglectful one, will influence a person's character and is liable to be a strong predictor of life outcomes. In this sense you could say that psychology, or the scientific study of how the human mind is shaped by and behaves in a certain context, has been so influential over the course of the last century that we have incorporated its basic concepts into our understanding of ourselves. We are psychologically literate even if we have never studied psychology or had therapy. We say things like 'She's got issues' or 'He's got a lot of emotional baggage'. We understand the ideas of trauma and repression, of conflict-avoidance and emotional intelligence. We also tend to be very heavily invested in the idea that an individual can 'work on themselves' to change aspects of their behaviour that they find undesirable. Even if they have suffered a miserable childhood or a tragic life event, perhaps especially if they have suffered in such a way, we want to believe that they can escape their past and reinvent themselves. Many of us know people who, through force of character or sheer will, have done precisely that.

Neuroscience is now presenting us with the opportunity to understand more about how such resilience operates and how we 'make our own luck' through exercising choices about which environments or people to cultivate. Those choices, taken as an adult, are informed by endless loops of interaction between previous experiences and our perception of the world. At the nub of it all sits our brain, the physical matter that we were born with inside our head, without which there would be no perception, no memory, no mind. Our brain develops in response to lived experience and changes throughout our lifetime, but a newborn baby's brain has already laid down foundations in the form of neural pathways that will shape the way the person interacts with the world for the rest of their life. There is something fundamental to our individual being besides the story we have created about ourselves. That something is an organ of such staggering sophistication and power that it is only now beginning to yield its secrets to science.

Over the past two decades there has been an explosion of study of this previously inaccessible realm, driven by massive technological advance. That study – the discipline of neuroscience – is illuminating the question of whether we are in control of our outcomes or destined at birth to follow a particular path by (sometimes literally) shining a light into the brain's deepest regions. It turns out that there is still some power in the old idea of fate, though not in the sense the Ancient Greeks understood it, as an external force. In its twenty-first-century incarnation, our fate is buried within our physical selves, in the hard-wiring of our brains and our genetic inheritance. A straightforward (if devastating) example of biology as fate would be carrying the genetic mutation for Huntington's disease, where individuals carrying a single genetic change will eventually develop problems with coordination, reasoning, flexibility in thinking, decision making and, in some cases, psychosis. A more complex manifestation is the highly nuanced way that we as individuals are predisposed to certain behaviours rather than others.

Is it possible to say that the brain we are born with determines our personality, beliefs or particular life events? This is the sense of fate that I set out to investigate. The central question we will be pursuing throughout the book is one of agency. To what extent are we in control of what we do and of what happens to us? How much of what makes us who we are is inherited at birth, written into the workings of our brain or flowing through our veins?

What exactly do I mean by fate and free will?

The dualism underpinning the concepts of brain and mind, biology and psychology, nature and nurture, fate and free will is artificial and only useful up to a point. There could be no life story without the brain to create it, and it seems to be the case that our brains are driven to create our unique narratives. Psychologists have for the most part given up asking the old either/or question 'Was it nature or nurture?' in favour of embracing the fact that the answer is always 'Both.' The eminent biologist Robert Sapolsky put it succinctly when he wrote in his book *Behave* that 'It actually makes no sense to distinguish between aspects of a behavior that are "biological" and those that would be described as, say, "psychological" or "cultural". Utterly intertwined.' People across all the cognitive sciences, in philosophy and psychology, those working in artificial intelligence, psychiatry or neuroscience, increasingly emphasise that when it comes to brains and their dizzying array of activities and outputs, there can be only a multi-pronged approach to increasing our understanding.

Given that I am a biologist – a neuropsychiatrist by specialism – my approach is inevitably shaped primarily by the discipline of biology. My aim is to investigate whether our fate can be understood in biological terms, though fate is perhaps too loaded a word for most of the outcomes I'm interested in, since it conjures up connotations of a tragic end. I'm looking at how we construct our individual sense of the reality of the world and how this impacts on our decision-making, which then compounds into behaviour and accrues into the material of our selfhood and daily life. Given that we're investigating brains in a context of biological determinism, though, I will also be discussing health outcomes, and mental-health outcomes in particular. So, I will be looking at fate from a number of different perspectives, through debilitating conditions, such as schizophrenia on the one hand, and a range of behaviours that impact on everybody's daily life on the other.

For some unfortunate people their biology truly is their destiny, but most of the time biology is not that simple in its operations of cause and effect. Biological mechanisms *contribute* to most disorders of the brain: they don't cause them in a straightforward way. For example, some studies cite that around 80 per cent of a person's risk for developing schizophrenia is down to the genes they were born with, but at current reckoning around 180 genes are implicated, and the way that they interact with each other and with the person's environment is yet to be fully untangled. When it comes to behaviours like food choice, friendship style, an aspect of personality such as sociability, or our beliefs, the biological mechanisms that contribute are vastly nuanced and interact in subtle ways with each other and with environmental factors. Which is not to say that an individual's choices and behaviours in these areas aren't predetermined by innate biological factors outside his or her conscious control. It just means that the idea of fate might need to be relaxed from its total and tragic connotations and understood as the destination we were always overwhelmingly likely to arrive at.

Throughout the book I shall consider the relative influence of innate factors such as our unique genetic inheritance or the evolutionary pressure that has shaped the human brain's physiology, and weigh them against the influence of learned behaviours shaped by environmental exposure. I use the terms 'innate' and 'learned' in the knowledge that, although it can be useful to observe and describe a behaviour as if under a microscope, it will only reveal its full self if it is studied in the round, like a jewel held this way and that to the light.

A biological approach that seeks to understand anything of the staggering complexity of human behaviour must itself embrace a multifaceted method if it wants to contribute insights to wider debate. So, even if I wanted to write a purely biological study of how (and whether) we consciously shape our lives, it would still need to take on board findings from biology's many different branches. I would have to consider chemistry, hormones, the prenatal environment, our genetic inheritance, our very early-years' experiences, epigenetics and evolutionary pressures. In other words, all biology is complex, and brain biology is at the more complex end of the spectrum.

To say something that may be useful to a non-specialist as they seek to understand the influence of neurobiology on their own life, or the lives of others around them, I have simplified arguments and tried to focus on real-life examples. My aim has been to pick my way through an endlessly branching labyrinth of fascinating research and emerging information on a pathway towards an idea that began to emerge for me some years ago. Neuroscience has made staggering advances in exploring how the brain produces behaviour and life outcomes, but the logical conclusion of these advances, that neurobiology determines our lives far more than we know or in some cases like to admit, has not yet been widely discussed.

We start with some fundamentals of brain biology by looking at relatively basic behaviours, such as what we choose to eat and whom we choose to have sex with. We move on via a discussion of how love, friendship and social structures are driven by neurobiology to how our brain develops and learns throughout our lifetime, then look at increasingly higher-level functions, such as how perception arises and how we form our beliefs about the world and our moral opinions.

The final chapters explore some of the pragmatic and ethical challenges that arise for both an individual and society as a whole from these discoveries. How can neuroscience's understanding of biological fate be applied, for example, to help thwart the destiny of people suffering with mental health and neurological conditions? If we can predict who will develop schizophrenia, or autism, addiction, depression, anxiety, mania or ADHD, is it right to intervene to 'improve' an individual's outcome? Which are the cutting-edge neurotechnologies that will shape our reality over the next couple of decades? In the future, could (and should) we all be taking neuroprotective treatments tailored specifically for our genetic brain weaknesses? And how can we work out which of our own traits is amenable to change and which simply need to be managed to reduce their potentially negative impact on our lives?

Since I very much wanted the book to participate in a conversation that reached beyond biologists and their insights, however powerful those insights are, I've spoken to people from all over the world who are working on understanding different aspects of how the brain creates our sense of self and determines our lives. I asked all of them not just about their own particular work but for their opinion on fate and free will. I knew the book would benefit from taking in the perspective of Christian theologians, social and evolutionary psychologists and Buddhist psychiatrists, as well as many of my fellow neuroscientists. Everyone was unfailingly generous and patient with me and I have drawn heavily on our discussions. They were also unanimously excited by the way the field of cognitive science is opening up, driven by exponentially improving technology and a staggering run of discoveries in the field of neuroscience. Their interpretations about what those discoveries mean varied, their opinions on how to apply them varied hugely, but the excitement was absolutely consistent.

A golden age of brain science

It's no exaggeration to say that we are living in the era of the brain. Until just a decade ago the human brain was considered an enigmatic structure of unfathomable intricacy. Billions of cells intertwined with trillions of connections to form the most convoluted and interconnected network imaginable. But now technological advances are providing novel ways of unravelling the make-up of our circuit board of thought. We can map and (within certain contexts) *control* thinking. We are able to observe how the brain operates – with high resolution and in real time – in fully conscious, moving, learning mammals. We can observe the brain's architecture and operations, see the birth of new nerve cells in even the elderly mind, watch as fresh neural pathways are formed to support new circuits of thought. We can peer beneath the skull and see habits taking shape, observe skills being learned.

This plasticity – the brain's ability to alter at a physiological level throughout our lives – has led to some rather overplayed claims. It is tempting to deduce that the capacity for brain plasticity into old age translates to an ability to change our behaviours and outcomes throughout our lifetime. It is tempting to believe that we can mould our actions and thoughts in any way we want to. Tempting, but not quite correct. It seems to me that, collectively, we are buying into a seductive but simplistic idea of brain plasticity: that we can set out consciously to hone our brains, in much the same way as we do our muscles, to achieve anything we want. A growth mindset permeates society, advocating that our every goal or desire can be achieved. We are sold the concept of unlimited agency and capability, a vision of free will on steroids that rejects the idea of constraints, whether biological or socioeconomic. 'Dream it, be it' is not, from a neurobiological point of view, an entirely convincing slogan.

An opposing view is emerging, not only from neuroscience but also from psychologists such as Daniel Kahneman, the Nobel-Prize-winning author of *Thinking, Fast and Slow*. Rather than stressing the brain's undoubted capacity for plasticity, this approach highlights its hard-wired nature and propensity for cognitive bias and over-confidence in its own powers of judgement. This view is much more challenging to our cherished notions of personal autonomy. It suggests that many of our decisions are made not by our conscious mind but as a result of deep automatic processes at a subconscious level. Those processes are determined by the physiology that we're born with and shaped by our genetic inheritance. All of which means that we're not in conscious control to anything like the extent that most of us imagine.

How can we reconcile these two opposing views of our behaviour? First, and crucially, they are not mutually exclusive. They are both valid and both 'true' in different situations and to different degrees, depending on what specific aspect of behaviour or which life outcome we're looking at. The 'cause' of any human behaviour is multifactorial – composed of many contributing factors rather than a single one.

Something as simple as choosing your lunch depends on a staggering number of factors. Your brain typically weighs just two per cent of your body mass but it consumes 20 per cent of your daily calorie intake. It's no surprise that this hungry beast dictates your food choices. As well as your innate preference for high-calorie, high-sugar and salty foods (we'll be coming on to humanity's appetite for these foods in the next chapter), there is also your individual capacity for mindful choice and delayed gratification, not to mention a whole host of eating habits and preferences built up over a lifetime. And that's just the back story. In the moment, your brain will be fielding incoming signals in the café that might influence you at a subconscious level. Your hormone levels that day exert an influence, as does how tired you are and whether you're coming down with a viral infection. Even with the choice of sandwich, decision-making is complex as well as largely unconscious.

When it comes to the bigger questions, such as who to marry or your opinion on the existence of God, the cognitive processes are exponentially more complex since they are carried out over a much longer timescale and draw on even more regions of the brain.

So, this is a big subject and there are no simple answers but there is an emerging body of scientific knowledge that reinvigorates an unfashionable and potentially uncomfortable view of human behaviour as driven and, to some extent, determined by innate neurobiological factors. It is impossible to say that any single action, decision or outcome was fated for us by our genes or hard-wired into our brain, but it is possible to say that someone is *predisposed* to take certain decisions due to the way their brain was constructed prior to their birth and the genetic inheritance informing its operation over the person's lifetime. A complex dance is under way between brain circuitry, deep biological drives and learned experience every time you make a decision, however seemingly trivial. And in the end a great deal of what we think of as unique to our life story – our dreams, fears, beliefs and loves – comes down to the millions of decisions that make up everyday behaviour, which in turn makes up our life choices and personality.

All of which raises numerous questions. Are some of our personality traits and behaviours fixed and others malleable? If that is the case, how, exactly, do we identify which is which?

What can we as individuals do about any of this? And to what extent is any of it true for all of us?

Avoid the neurohype

I feel fortunate to be living through the age of the brain and I believe that neuroscience can make a robust contribution to answering the questions humans have always asked about themselves. I wouldn't have written a book about the science of fate if that were not the case. But neuroscience on its own is not the answer to life, the universe and everything. Some critics view it as a reductionist discipline that overemphasises the brain (or, even worse, the brain *scan*) at the expense of a more holistic approach to psychology and social and cultural life. In fact, some of the most fascinating work currently being undertaken by neuroscientists looks at the brain as part of a holistic network that takes direction from the gut and our immune system, as well as the signals from our environment.

I have tried to embed my discussion of neuroscience's insights within a broader context and resist too much flattening out of the complexities of human behaviour. Sally Satel and Scott O. Lilienfeld put their finger perfectly on the neurosceptical mood when they called their 2015 book *Brainwashed: the seductive appeal of mindless neuroscience*. It was precisely the overenthusiastic and sometimes misinformed use of concepts such as plasticity that sparked my determination to make a case for innate neurobiological factors being fundamental. Neuroscepticism is valid because, all too often, popular neuroscience behaves as if biology were not complicated. Pseudoscience and biological essentialism are the result.

It is obviously too simplistic to say that a brain scan *on its own* can 'prove' much about the complexities of an individual's mind but that does not mean neuroscience is nothing but hype. Between 2011 and 2012 the Royal Society published the results of its investigation into developments in neuroscience, their implications for society and public policy. The report is careful and considered but, ultimately, it endorses the idea that when neuroscience recognises that 'Each person constitutes an intricate system operating at neural, cognitive and social levels, with multiple interactions taking place between those processes and levels', it has earned its right to be taken seriously as a key component of this system.

How I got hooked on brain science

My own fascination with the endlessly interesting human brain came out of working with people suffering from psychiatric disorders. I was intrigued by the question of resilience – why some people move on from a seriously negative life event while others struggle to recover. During the late 1990s I was a nursing assistant at one of the UK's leading psychiatric hospitals, where I worked with children aged from twelve to eighteen who had been detained under the Mental Health Act. They were sectioned and placed in a secure institution in an attempt to protect them and others. Most of the patients had been sent there from across the UK after numerous unsuccessful attempts by their own local health authorities to support them. The majority had experienced abuse or neglect early in life. They were extremely vulnerable to peer pressure and found it difficult to lead a healthy, happy life in the outside world. Their destructive behaviours included self-harm, drug abuse and hurting others, and they had varying diagnoses, ranging from schizophrenia, personality disorder and severe autism to bipolar disorder. A high number had criminal records for offences ranging from petty theft and mild antisocial behaviour to the more troubling, such as bestiality. I worked in the hospital on and off for three years, before I started studying biology at university and then during the holidays and at weekends.

I have many positive memories of the place. I remember patients playing basketball in the courtyard, enthusiastically drumming the bongos during music sessions in the living room, playing hopscotch in the corridors or quietly reading Harry Potter novels in their bedrooms. But my overriding memory of the experience is a feeling of claustrophobia and frustration for the children. I remember the heavy double-locking doors, the stuffiness of the wards, the lingering stench of the dense canteen food, the battle with medication-induced lethargy, the patients' constant preference to be sunk in the sofa watching television or snoozing during the day. For years I tried, along with the rest of their therapeutic team, to help them. The truth is, though, that in most cases I saw little improvement in their symptoms. The whole experience created a deep desire to contribute to the search for more effective help for these people.

It also left me with questions about what makes us, well, *us*. Many of the staff working in the hospital had experienced similar upbringings and challenges in life but were able to go home after their thirteen-hour shifts, unlike the sectioned patients. Why was that? What were the underlying differences that produced such divergence in life's trajectory? Could anything be done to help bulk up a person's self-protective abilities so that they could flourish no matter what life threw at them?

After my undergraduate degree in biology I went on to do a PhD in neuropsychiatry at Cambridge University. I joined a growing body of researchers working to understand the nuts and bolts of what makes us think and behave as we do. I wanted to bring together in this book what I have learned about the extent to which many behaviours are innate and decisions are taken at a subconscious level with what neuroscience was also revealing about the brain's capacity for growth and change. It has been a fascinating journey into understanding a little more about the factors that contribute to shape our behaviours and direct our life's outcomes.

Embracing the idea of biological fate

The science that suggests we are all, to a large extent, at the mercy of our neurobiology, driven in the direction of certain decisions and behaviours, susceptible to certain conditions, is very compelling. On one level every one of us, however uniquely complex and valuable, is also simply a human animal whose principal purpose (as we will see in later chapters) is to interact with others to exchange information that will contribute to the collective consciousness and, if we're lucky, pass on our genetic material. Deep drives are at work to further those basic goals and they are largely beyond our control.

Even what we think of as the more individuated aspects of our behaviours, the ones that we feel instinctively must be the product of nurture more than nature and more under our own conscious control, are formed at a deep level by innate factors we were born with and that were reinforced in our earliest years. Our personality, our beliefs about ourselves and the way the world works, how we respond in a crisis, our attitude to love, risk, parenting and the afterlife: any of the highly abstract opinions and character traits you care to mention are deeply shaped by how our brain processes the information it receives from the world. When we start to probe the idea of being a free agent in control of our life in the light of what neuroscience is now showing us, it can feel as if the space available for free will is shrinking fast and we're stuck in a loop that refers us back endlessly to a prior stage of preordained experience.

The incredible boom in neuroscience over the past two decades means that we are living in a new age of scientific discovery. It is still early days but eventually its impact is likely to be as profound as that of Darwin's Theory of Evolution or the development of the laws of quantum physics. Over the next decade I expect to see more and more breakthroughs in the treatment available for people like my former psychiatric patients, as well as individually tailored support for those of us living with anxiety or depression. The concept of biological fate at its most deterministic – a genetic mutation that dictates the development of Parkinson's, for example – will soon be overturned by new treatments on the horizon that enable a scientist to turn off that mutation at the flick of a genetic 'switch' or a surgeon to correct the erroneous brain circuitry using the power of electricity.

During my lifetime there will be significant discoveries, applications and ramifications. It's possible that, as we discover more about the neurobiology of belief formation and prejudice, we might be able to boost our openness to new ideas, say, with massive consequences for reducing conflict at every level.

Not that it will be straightforward. Our predecessors were shaken to the core by the ideas of Newton, Darwin and Einstein. They had to re-evaluate humanity's place in the universe. Perhaps neuroscience is now demanding of us that we embark on a similar journey of thought disruption. We as a society will certainly have to consider the implications and ethics of its insights. On one (relatively straightforward) level we'll need to decide collectively whether treatments should be developed for genetically inherited conditions and how to make sure they don't become a luxury of the rich.

But there are even more challenging questions. If free will does indeed occupy a shrinking space in our increasingly well-mapped brains, we will have to perform some serious mental gymnastics to work out how we feel about that. Any move to suggest that we are less in control of our own lives than we imagine comes with risks. At an individual level it can be not just uncomfortable but also destabilising. People who believe their actions have no impact on a situation tend to feel disempowered and behave in less socially responsible ways. The impact on society of us all relinquishing our belief that we are in control of our destiny might be catastrophic.

Could neuroscience provide a framework for understanding our behaviour that foregrounds the proven biological influences without diminishing an individual's sense of validity and interconnectedness? Could it develop a convincing argument that, though individually we are less in control than we thought, needn't condemn us to selfish individualism? I believe it can and will. The emerging neuroscience of compassion substantiates the idea that the concept of humanity's innate selfishness has been overplayed. It is just as possible to argue that we are predisposed to value our social interactions and behave altruistically.

All these questions depend on science that is still in its infancy. In the meantime, perhaps we can accept that even if free will is an illusion it is a necessary one. Going back to our earlier point about how each one of us inhabits our own mind as if it were the universe, then the version of reality we have constructed for ourselves is inescapable, even if it's illusory. Robert Sapolsky, an enthusiast for biological determinism whose rejection of free will is, in theory, unshakeable, says, 'I can't really imagine how to live your life as if there is no free will. It may never be possible to view ourselves as the sum of our biology.'

We shouldn't jettison our deeply held belief in our own powers just yet, but a fuller understanding of their limitations is essential if we're going to have a debate about how neuroscientific knowledge should be applied. From NHS priorities to bioethics and the future of education and public health, our society will alter under the impact of what's being uncovered about the functioning of our brains. For us as individuals, knowing more about how neurobiology drives behaviour puts us in a better position to take those decisions over which we *do* have control. As we'll see in Chapter 3, for many people it's easier to choose to eat well once they know how circuits in their brain respond to and control appetite (for food, sex, attention and almost anything else you care to mention).

The era of the brain is one of the most exciting times I can imagine to be alive. It's now possible to admire the elegance and sophistication of humanity's processing system as it goes about its work. These new vistas need not reduce our appreciation for what it means to be human. Instead, they offer the opportunity to marvel at how the full range of human behaviour is produced from such an intricate yet ultimately simple design.

That sense of wonder can be extended to take in not just our own brain's magnificent achievements but those of the collective consciousness of the seven billion or so brains on the planet, each with the 86 billion nerve cells and 100 trillion connections that make up the circuit board of each individual mind. From this staggering network of interconnected processing power arises a collective, species-wide experience that drives far-reaching evolutionary change, and creates an infinite variety of human stories. We are all destined to be part of humanity's creative process of development.

Knowledge is power, as I had to remind myself when I was feeling jittery and calling up my surgery for the test results. The more we understand how our brains, bodies and the environment work together, the more every one of us can contribute to the neuroscientific revolution now under way. Time to begin at the beginning, then, with a look at the brain we are born with, and how it develops over the course of our lifetime.

CHAPTER TWO

The Developing Brain

On the day a baby is born her brain is already a marvel of accomplishment and fizzing with the potential to know and do even more. A newborn may be utterly dependent on her caregivers but she is also capable of interaction and rudimentary communication with them. She is primed to explore her environment to learn so that she can, one day, cater for herself. Babies are bundles of curiosity and raw emotion, pure will and deep social instincts, poised at the beginning of a lifelong quest to discover more about the world around them.

Having my own child has made me see this process and the extraordinary achievements of the developing brain as little less than miraculous. It's one thing to read the textbook account of an infant's brain regions gradually connecting up to allow behaviours to emerge, and another to see my son's consciousness form in this way. Admittedly, when he's in the middle of a tantrum I sometimes catch myself willing his prefrontal cortex and language circuits to hurry up and get online with the rest of his developing brain, knowing that until that happens he'll be unable even to *begin* to learn to regulate his emotions or communicate his needs somewhat more politely. But most of the time I just marvel at him.

Here he is: another beneficiary of the incredibly sophisticated organ that is the human brain.

Each of us develops at different rates, as the health visitor reminded me recently during my son's routine check-up. Even as we progress from baby to infant, from teenager to adult and beyond, each person is unique, the culmination of all their experiences. There is scope for a magnificent variety of behaviour in each of us. This, in large part, is due to the complexity of the human brain. Its mesmerisingly sophisticated and ever-changing landscape is responsible for the myriad complex emotions, thoughts and behaviours for which each of us has the potential. So it is problematic to suggest that there is any such thing as a 'typical' or 'average' brain, responsible for a 'typical, average' human life. But to try to get to grips with how our individuality, personality and unique life decisions are made, we must start by looking for patterns and making generalisations. We will be drawing on the huge body of research into the way the brain's structures and functions generally alter over a lifetime, but it is crucial to remember that those changes are all shaped by our particular circumstances. Our unique inherited genetic package, as well as the familial and social context in which we find ourselves, delivers a highly nuanced version of brain change for each of us. From the standard developmental stages arise billions of unique brains, and the foundation for every person's life story.

In this chapter we will be looking in detail at how our brains function, how we learn, and about how those processes generate what we think of as our 'selves'. We will assess how a baby ensures their every whim is catered to; why toddlers throw those maddening tantrums; examine stroppy and impulsive adolescents; and uncover why accumulating knowledge leads to the profile of an older brain, exploring the neural basis of wisdom and, on the flip side, closed-mindedness. We'll consider why the human brain often becomes so frail in old age and what we can do to help maintain its capabilities for as long as possible. This chapter will allow us to consider the infinitely complex interplay of innate and environmental factors on the development of behaviour in a familiar context: that of a typical human lifespan. With these principles in place we'll be ready to examine whether our choices are predetermined in the context of specific behaviours, from what we eat and who we have sex with to how we form our beliefs.

Let's start at the beginning with a look at the brain of a newborn baby. After all, that's the conventional beginning for most life stories, one fabulous exception being Laurence Sterne's Tristram Shandy, who opens his tale on the night of his conception and takes several pages to get past it. In biological terms, of course, he's quite right. The story of a human being begins way before birth, since a baby's brain has been developing during the previous nine months of gestation, shaped by everything from evolutionary pressures to genetics, what the mother eats to what the paternal grandfather ate. But let's consider the brain our baby is born with, and what happens next.

We are probably all aware that a child's first few years are crucial in terms of influencing their life outcomes. It's an incredibly dynamic period for cognitive development. Specialists in disciplines from psychology to linguistics have produced decades' worth of research showing that the influence of environment and experience during a child's earliest years can have effects that last a lifetime, for good or ill. And there is a sound anatomical reason for this. Although the brain's building blocks for information-processing – the neurons, or nerve cells – are mainly constructed while the baby is still in the womb, the complex process of connecting everything up largely occurs during its first three or so years.

The brain of a child born at full term contains around the same number of neurons as an adult's, although it takes up only about 25 per cent of the volume of the adult brain. By the time the child reaches her third birthday, her brain has developed to, on average, 80 per cent of its adult size. Each of the nerve cells has increased in volume, branching out to initiate extensive and intricate connections with other cells. These connecting structures, which look like branches extending from the trunk of a tree when viewed under a microscope, are called arborisations, and the gaps between them and the next neuron(s) are called synapses. One particular kind of arborisation, the axonal ending, produces neurotransmitters that carry the electrical signal conveying information across the synapse, from one cell to another. During the child's first three years the synapses form at a faster rate than at any other point during our lifespan, creating the basis for the connectome, or the circuit board of our mind. The circuit board determines how information from the outside world is processed, and shapes our behavioural responses. So, the early brain-sculpting process literally dictates how the baby will go on to see the world and interact with it as an adult. No wonder parental anxiety has exploded in the Age of the Brain.

Fated at birth?

Nothing sums up the sensitivities of the interaction between innate characteristics and environment better than early-years development. Babies are born with some highly complex wiring already in place but the influence of environment in shaping the developing connectome is also uniquely important during the early period. Most of us can't resist seeing a baby as a blank slate, pure latent potential, but it isn't quite that simple.

To explore the impact on brain development of the formative years of a person's life, I met Dr Victoria Leong, who heads the Baby LINC (Learning through Interpersonal Neural Communication) Laboratory at Cambridge University. I particularly wanted to discover more about the interplay between a baby's innate brain functions and the influence of environmental inputs during this super-sensitive early period.

Vicky has been studying how infants develop for almost a decade and is a fount of wisdom and reassurance for any parent who has ever worried about how to support their child in the crucial early years (and surely that's every parent). I had a lot of questions for her but the first thing I wanted to establish was a clearer picture of the skills and capacities that the baby emerges with, fresh from the womb.

It turns out that newborns are already socially capable and highly curious. The drives to form social bonds and to explore the world are both implicated in all sorts of adult behaviours, from the basics of reproducing, making friends and defining social groups to developing belief systems. They are provided to us even before birth.

'If you look at a baby's typical behaviours at just a minute old, it's all about keeping close to the caregiver,' Vicky told me. 'The sucking and grasping reflexes facilitate bonding. Babies are also born with a desire to learn from their social environment. They want to engage and understand other people by, for example, maintaining direct eye contact, looking at faces, sticking out their tongue in imitation when somebody else does. They act in ways that are likely to continue the interaction with the adult. It's as if they're trying to understand the social processes behind that action.'

There is an obvious benefit to this behaviour for the individual, of course. Looking at it rather clinically, those social skills undoubtedly help to charm the caregiver, locking them into their new and vital role of meeting the baby's every need. It makes sense that a newborn is socially responsive, thereby upping the adorable stakes and ensuring that he or she will receive the help they need to survive the first critical period of development, while the rest of their nervous system wires up.

It's this 'wiring up' of existing brain structures that gives rise to the huge developmental leaps that babies and small children make. Different domains of the brain have specific sensitive periods for learning different skills, when new connections are laid down extraordinarily swiftly. Small amounts of what's called 'pruning' take place in tandem, as experience dictates which circuits to keep and which to chuck out. It can feel, to an observing adult, as if suddenly another piece of kit comes on board and brand new behaviours emerge overnight. Needless to say, it's never quite as simple as a neural pathway connecting up and then, snap, you get a behaviour, but the metaphors of 'pennies dropping' or 'something clicking into place', which describe our amazement at watching a child do something for the first time, have some validity. I asked Vicky to talk me through the essentials of her specialism, language acquisition, to focus on the process by which anatomical change in the brain supports and enables behaviour.

Vicky studies the connection between language acquisition and the hearing system. They are intimately linked and their development illustrates the way that babies are born with innate skills that are then fine-tuned to suit their particular environment. All babies who are not hearing-impaired are born with a mature cochlea that allows them to assess pitch and loudness. They are also born 'citizens of the entire linguistic world', able to hear phonemes (speech sounds, like *p* or *s*) that are used in any language spoken across the globe, and to discriminate between them. But as a baby is exposed to its native language(s), he or she loses the ability to hear phonemes that don't occur in their environment. For reasons of efficiency, the baby's brain tunes into the speech sounds that are directly relevant to them. Without this ability to screen out background noise and focus on what is linguistically crucial it would be impossible to learn first to understand and then to produce speech.

As Vicky explained, 'If, for example, you test a ten-month-old American infant, they have literally stopped hearing phoneme discriminations that are used only in, say, Hindi. This is part of what we call "perceptual tuning". In order to do this, brain changes have to take place. The auditory cortex, the part of the brain close to your ears, needs time to mature. The nerve circuits here are connected up and refined based on your environmental experience, and this process takes roughly the first year or so of life.'

The experiences to which we are regularly exposed regulate our perception, priming our brain to receive only those aspects of information it deems important, then to filter and discard the rest. Over time this not only shapes how you sense the world but also instructs how you interact with it. The acquisition of language is obviously dependent on input from the environment, but even the simplest behaviours derive from highly complex interactions between neurobiology and lived experience. So a child discovers, for example, that bright sunshine streaming directly into her eyes produces a sensation of discomfort. She will learn, through repeated exposure, to associate that discomfort with the source of light and will then engage circuits across the whole brain, including the problem-solving prefrontal cortex, the memory-retaining hippocampus and the movement-generating motor cortex to respond to it. All this brain activity allows the child to roll away from the light. Or it sends electrical impulses along the nerves in the body to instruct her arm muscles to reach out and pull a blanket over her eyes, or, when she's a little older, either to move away from the light source, remember where her sunhat is located and put it on or summon an adult and communicate her desire for the curtain to be closed. Whatever the solution she opts for, previous experiences and observations have facilitated the wiring up of regions in the body and brain to allow this to happen, in an extraordinarily complex but elegant process.

Babies' brains are constantly busy and the outputs of all their hard work, from increased physical capability to new social skills, emerge thick and fast. There is a typical developmental model that can describe the order in which milestones are likely to be achieved, but as any baby book worth its salt will tell you, there is tremendous variation between individuals. As a general rule, the greater the number of different brain regions involved in producing any particular behaviour, the more complex that behaviour is likely to be and the more scope for variation in its emergence.

Learning to control emotion is self-evidently a complex piece of behaviour if the difficulty of its acquisition is anything to go by. Tantrums are an all but inevitable part of toddlerhood because to be able to control strong emotions, such as frustration, jealousy or anger, a child needs to integrate information from the areas of the brain involved in the generation of emotion, language and reasoning to articulate her emotions, take account of others' feelings and respond to all this information in a positive way. This starts to happen around the third birthday, give or take a few months, and can take some time to finesse. Until then the disparate regions are not fully connected to one another, which makes meltdowns a fact of life. Even once the neural circuits have been laid down and the brain regions are in touch with one another, it takes more time for those circuits to be strengthened, through experience, to the degree that a child can reliably exhibit emotional control.

This illustrates something crucial about the learning process and its biological mechanisms, which holds true throughout our lives. As a new skill is practised or we repeat a realisation, the neural connections that support it are strengthened so that the learning is consolidated into a memory. If the memory is repeatedly visited, it will become the default route for electrical signals in the brain. In this way learned behaviour becomes habit. Neural connections that are not used are eventually lost, via pruning. Most connections between nerve cells occur on minuscule structures termed 'dendritic spines' that change shape in response to electrical activity. As learning occurs, dendritic spines reach out to make contact with the neighbouring active nerve cell. The dendritic spine swells and eventually splits into two daughter spines, thereby doubling the circuit connections. This is the process by which each nerve cell can connect to up to 10,000 others, giving rise to the roughly 100 trillion connections collectively known as the connectome. That's the mechanics. But it's the gradual process of taking in signals from the environment, observing other people's actions and seeing patterns in reactions, that allows our brain to create our individual connectome, which enables each of us to lead our unique life.

Now, there's a plethora of neuroscientific research out there trying to unravel whether caregivers can do anything that might boost this process of connecting up the nerve cells in the most positive fashion to provide children with the best start in life. It gets turned into advice about parenting that ranges from the robustly scientific and sympathetic to the downright unhelpful. As a first-time mother, and despite my doctorate in neuroscience, I found it overwhelming trying to wade through all the literature on the subject. In the end I stopped trying to distinguish between the good stuff and the spurious claims that tap into new parents' anxieties and simply went with my gut instinct. In retrospect, though, should I have done things differently?

With some trepidation, I asked Vicky if her work suggested there were any tangible actions a parent can take to help a baby's brain wire up in the best possible way. Focusing on her specialist area, she told me that the quantity and quality of speech to which you expose your baby are important. Her general advice was simple: talk as much as possible to your baby. It gives the child more material to learn from. That's the quantity part.

When it comes to quality it's not necessary to read Shakespeare aloud over your baby's cot. Quite the opposite. Caregivers from different cultural and linguistic contexts seem to be pre-wired to talk to babies in what is termed 'motherese', that singsong cooing on a slightly raised pitch that parents adopt and to which babies pay particular attention. The phenomenon was identified in experiments more than fifty years ago. Primary caregivers of either sex use this speech style spontaneously, though mothers typically do it more than fathers. (This may reflect that women have until very recently been the primary caregivers of their own and other people's babies.) Such 'infant-directed speech', to give it its more technical name, seems to play a part not only in language acquisition but also in boosting a baby's ability to pay attention and manage emotion. Essentially, language gives toddlers the tools they need to learn to soothe and express themselves.

Vicky also told me about her groundbreaking research into the astoundingly positive effect of direct eye contact between babies and caregivers, which enhances brainwave synchronisation between them and spurs the babies' efforts to communicate. Looking straight at your baby while you talk to them boosts their rate of learning. Vicky discovered this by getting baby and parent volunteers to wear electroencephalogram (EEG) hats. Embedded into the hats are hundreds of electrodes that pick up electrical activity in the brain. This enables scientists to read the brain waves emitted by nerve cells as they communicate with each other to generate our thoughts and emotions, and direct our movements. Vicky and her team measured electrical activity in the brain while the parent and baby pair interacted in different ways.

Now, neurons operate with oscillations of activity, firing at specific times. We experience our vision as a continuous stream of images, as if from a video camera, but in fact the eye is taking snapshots. The brain then processes these samples to generate our smooth perception of the world. It turns out that simply looking the baby in the eye while talking to them helps to synchronise brainwave patterns so that baby and parent are filtering information, and literally seeing the world, in a similar way. Vicky's work demonstrated that the boost to the babies' language acquisition was remarkable.

This synchronising of brain waves between individuals can also be harnessed later in life. Direct eye contact with native speakers can be crucial in mastering a new language as an adult: it seems to be a mechanism for reopening sensitivity to foreign sounds. Interestingly, exposure to the target language via television is not sufficient. It's the live feedback loop of synchronised brain waves that makes the difference. These findings have important implications for learning, particularly in our increasingly digital world, and their impact is not confined to language acquisition. When adults sing in a choir or discuss an issue in a group, their cohesion is remarkably improved with eye contact.

After my interview with Vicky, I made an extra effort that evening to engage with my son. The TV was switched off, we went to the park and I might have concentrated on direct eye contact a little more than is natural. He made a polite request that I stop staring at him.

Vicky's studies, alongside many others, contribute to the mounting stack of evidence showing how early experiences sculpt the way in which an individual sees, hears and perceives the world, with enormous implications for behaviour. There is scope to alter the brain's circuit board later in life but effort is required to do so. And if we consider that a child's early-years environment is largely dictated by her caregivers, who (assuming they are the child's biological parents) have a similar genetic complement and will experience the world based on their own early-years experiences, we start to see how behaviours can be perpetuated across generations. What we think of as basic aspects of our highly individual personality have been doubly determined, many years before, by factors outside our control.

The neuroscience of personality is a very new field, but it is generating impressively robust results at a cracking rate. A fascinating 2018 study was notable for its debunking of one of the all-time-classic behavioural psychology tests relating to personality formation. The 'marshmallow test' has all but passed into folklore. It's certainly been cited in dozens of parenting books. A series of studies was conducted in the late 1960s at Stanford University to examine whether you can predict a child's achievement trajectory from a young age. The researchers were particularly interested in delayed gratification. Six hundred children, aged around four and a half, were offered a choice between one small reward, a marshmallow, provided then and there, or two marshmallows if they waited fifteen minutes for them. When the researchers followed up around twelve years later, they found that the children who had been able to delay immediate gratification now exhibited advanced traits of intelligence and achievement compared with those who had caved in to temptation. It seemed that the ability to exert cognitive control over impulse and appetite-driven behaviour from an early age functioned as a predictor for how your life might shape up.

A team of neuroscientists from New York and California universities set out to see whether they could replicate these results. Instead they found that any difference in attainment between the impulsive and resolute four-year-olds had largely disappeared by the time the children were fifteen, once their parents' or carers' socioeconomic background and education were accounted for. Children from affluent professional families were generally doing better than peers from less comfortable backgrounds at fifteen, regardless of their behaviour at four.

It seems that the Stanford researchers had omitted to factor these dimensions into their experiment design. The new results make intuitive sense: growing up in an environment of scarcity can lead people to opt for short- rather than long-term rewards; a second marshmallow seems irrelevant when a child has reason to believe that the first might vanish at any moment. If their parents are not always able to stick to promises because money runs out, or their siblings steal their treats, immediate gratification is a perfectly rational strategy.

This reads to me like a cautionary tale. All science, of whatever discipline, is provisional and subject to its authors' constraints and cognitive biases.

The salutary lesson of the marshmallow test should be borne in mind when we consider the next neuroscientific examination of the emergence of personality although, given that it reviews a vast amount of data from numerous different studies, it looks promisingly robust. The paper was published in 2005 by Avshalom Caspi, University of Wisconsin, Brent Roberts, University of Illinois and Rebecca Shiner, University of Colgate. It suggests that personality traits are fairly stable over the lifespan, so that on average the personality of a toddler, or even a younger baby, will tell you something about that of the adult later in life.

Personality traits are commonly assessed in terms of what's known as 'The Big Five' of extraversion/positive emotionality, neuroticism/negative emotionality, conscientiousness/constraint, agreeableness and openness-to-experience. There are some obvious difficulties in trying to assess a baby's reactions to scenarios and their general baseline temperament, and the study authors acknowledged this complexity when they concluded, 'Despite the challenges inherent in mapping out temperament and personality structure across the life course, researchers have made substantial progress in elaborating taxonomies of individual differences in both childhood and adulthood . . . Behavioural genetics research has uncovered increasingly reliable and robust evidence that genetic factors substantially influence personality traits.' Could you therefore screen babies, perhaps even before birth, to help predict their future personality and temperament?

The attempt to identify genes associated with complex conditions, traits and behaviours is extremely challenging, as we will see, since multiple genes of varying but small effect-sizes are implicated. It is not as simple as saying, 'Here's the gene for extraversion.' But if, as this review suggests, personality traits, as we currently understand them, are fairly stable over a lifetime, that provides a potentially valuable source of self-knowledge, particularly since we might be justified in viewing our personality as a foundation for making decisions in later life, from our career path and friendship groups to our hobbies and holiday preferences.

There is always a danger of becoming wedded to a fixed sense of our own personality, since it risks imprisoning us in a self-definition that may in fact be the product of a particular set of circumstances, or even straightforwardly deluded. As we will see in the next section, a great deal of the impulsive or reckless behaviour exhibited by teenagers, for example, is specific to their stage of brain development rather than to their individual personality. But it does increasingly appear that personality begins to emerge early in life, tends to remain stable and can act as an important tool that shapes our life trajectory.

Are teenagers the product of nature, or nurture?

Let's flash forward a few years and move from considering infancy to adolescence. You may remember Harry Enfield's infamous sketch, in which the clock strikes midnight, Kevin turns thirteen and instantly becomes the parody of a moody teenager, incapable of normal speech and embarrassed by everything his parents do. The stereotype feels extremely recognisable, but how much of Kevin's behaviour can be attributed to the biology of the adolescent brain and how much to social pressures? Is the moody teenager a cultural construct, the product of modern individualistic societies where adolescence is extended and indulged?

The answer seems to be no. Adolescence is empirically associated with novelty- and sensation-seeking behaviour, more extreme risk-taking, self-absorption and susceptibility to peer pressure. These traits have been observed across time periods and cultures. Socrates was disparaging about the youth of his day, saying, 'The children now love luxury. They have bad manners, contempt for authority, they show disrespect for elders in place of exercise.' Rousseau captured the narcissism of the typical teen: 'At sixteen, the adolescent knows about suffering because he himself has suffered, but he barely knows that other beings also suffer.' So although the word 'teenager' was only coined in the 1950s, it seems that the particularities of adolescent behaviour have been observed for millennia.

Kevin has numerous antecedents. In fact, his qualities are observable even in other species. 'Teenage' mice and rats will drink greater quantities of alcohol in one sitting than adults, and this age-related difference is accentuated when in the company of peers. Yes, you read that right: teenage rodents also hang about in gangs and get drunk together. Alcohol acts as powerfully on the hedonic circuits of a teenage rodent as it does on a teenage human, and the rodents, like the humans, are highly susceptible to peer pressure.

As we will see, the neurobiology of adolescence is absolutely key to determining stereotypically teenage behaviour, and is just as fascinating and just as dynamic as what happens during a baby's early years. Processes of development and the influence of hormones conspire to affect both brain and body and result in impulsivity, susceptibility to peer pressure and acute self-consciousness. One of the brain regions that undergoes significant changes throughout adolescence is the prefrontal cortex, directly behind your forehead and involved in a raft of higher cognitive functions, including decision-making, future-planning, inhibiting inappropriate behaviour, preventing us from taking unnecessary risks, understanding other people, so-called 'social cognition' and self-awareness. So, it's a pretty important brain region and, given that list of behaviours, its role in adolescent behaviour is likely to be crucial.

At the start of adolescence the brain already has in place some well-established neural highways within its networks but now, as well as continuing to create further connections, it also begins to do much more pruning of the less frequently used ones. The pruning occurs over our entire lifespan and is the basis for the perceptual tuning that Vicky mentioned in relation to language acquisition. But the rate at which it occurs appears to ramp up during adolescence. The teenage prefrontal cortex is the site of a vast amount of such synaptic pruning as it sets about refining what it has learned and building on past experiences at the same time. It has been suggested that during this highly dynamic period there is a mismatch between how information is processed in the prefrontal cortex and other deeper areas, including the reward circuit. The result is that an adolescent tends to be highly sensitive to immediate gratification and reward but hasn't yet fully developed their impulse control and decision-making. On average, they are more likely to act with immediate highs in mind than play it safe.

There's another important facet to adolescent brain development. During the teenage years the grey matter in the brain shrinks. In the prefrontal cortex it reduces by a whopping 17 per cent. Grey matter is a key part of our central nervous system. It's the mass dendritic branching where the synapses occur, plus nerve cell bodies and the accompanying supporting cells that form most of our brain and run down through our spinal cord. Pretty important, so losing 17 per cent sounds bad, until you realise that something is occurring to replace it. The loss can't be accounted for merely by all that pruning of redundant synapses. Some of the grey matter is replaced by an expansion in white matter, the generic name for the fatty lipids that wrap themselves around the nerve cells' long grey cylindrical structures, the axons. This coating helps to insulate them so that the electrical signal can be carried from neuron to neuron faster and with more integrity.

The various processes of teenage brain development taken together help to refine a child's connectome, upgrading it from a system composed of numerous branch lines to one based on fewer high-speed central lines. In essence, by the end of this critical time period (and experts increasingly agree that adolescence extends into our mid-twenties) information from the outside world can be processed with speed so that decisions can be made rapidly, honed by the benefit of experience.

Sarah-Jayne Blakemore, professor of cognitive neuroscience at University College, London, is one of the world's pre-eminent experts on the links between these brain changes and teenage behaviour. I first met Sarah-Jayne a decade ago. During the intervening years she's helped to invent a new field of neuroscience: studying how adolescence constitutes a distinct brain-development stage, different from either child or adult processes. She's also raising teenagers, so she's pretty experienced, whichever way you look at it.

Sarah-Jayne is adamant that we shouldn't demonise adolescence. She maintains that the adolescent brain is not a dysfunctional or defective adult brain; it's categorically different. Teenagers are living through a distinct and formative time of life when neural pathways are malleable and passion and creativity run high. Which is not to say that withdrawal from the family, recklessness, susceptibility to peer pressure and all the other commonly observed behaviours aren't real, and potentially problematic, but Sarah-Jayne's work has helped to demonstrate that there are good reasons why they occur. They can be boiled down to the need to form a more independent identity and learn how to operate outside familial contexts. It can sometimes be painful to be a teenager, or around a teenager, but it really is a vital learning period.

Neuroscientists have also studied the intersection of two issues that most worry the parents of teens – their tendency to take stupid risks and to be in thrall to the influence of their peers – by investigating why teens are more reckless drivers when their friends are present. As Sarah-Jayne has said, with the understatement typical of a research scientist, 'The need for social acceptance by one's peers plays a pivotal role in a lot of adolescent decision-making.' She's conducted experiments that demonstrate adolescent brains are hypersensitive to social exclusion, experiencing heightened anxiety and lower mood in the aftermath of a snub from a peer, in comparison with an adult. Teens are not wrong to worry about being rejected by a peer group, since friendship is a vital facet of wellbeing for all of us, irrespective of age. With the right tribes formed, friendships can protect against adversity in the future. So the needs of the developing social brain are at the forefront of a teenager's focus on their peers. And though teens are undoubtedly bad at assessing risk objectively (they are startlingly resistant to changing their assessments even when presented with credible statistical evidence that sways the vast majority of adults), Sarah-Jayne points out that taking risks isn't always a bad thing. It can lead to new experiences, learning and personal development. And it can be fun.

Adolescence may be the time when we are most self-conscious and most susceptible to peer pressure, but that's all part of building our unique identity, which is a key task of the teenage years. To investigate how this happens Sarah-Jayne's team ran an experiment in which they asked teenagers and adults to envisage and discuss their future. For both groups the so-called 'social-brain' circuit lit up with activity during the task but teens' brains became particularly active in the medial prefrontal cortex, while the adults' brains were active in different areas, more associated with memory. Sarah-Jayne believes that this points to how adolescents and adults use different cognitive strategies when thinking about themselves. Adolescents seem to have to concentrate to think about themselves in the future, and to compare themselves with their peers in order to do so.

But for adults, thinking about the self has become automatic and relies less on conscious thought. They have greater reservoirs of stored memories and experiences and can dip into these with ease when planning for the future or deciding how to respond in social situations.

But, of course, those experiences must be lived to create the memories that serve the adult as a manual for how to behave. Contact with people from outside the direct family group allows fresh vistas and ideas to influence a teenager's brain. And the teenage brain has evolved with the more primitive emotional and sensation-seeking brain regions in charge precisely to facilitate this. The impulsive, novelty-seeking teenager is essentially trying to build up a larger repertoire of experiences to help sculpt the prefrontal cortex in a unique way that will lay down their decision-making and thought processes for the future, while also attempting to construct the friendship tribe that best slots in with their individual reward-system preferences. And if that sounds exhausting, that's because it is. No wonder teenagers typically need a later morning wake-up call.

Sarah-Jayne is working with education and social-care policy-makers to help translate her work, and that of others in this new field, into the classroom and care settings. Adolescence, like a child's early years, is such a dynamic period of cognitive development that it offers a further opportunity to affect an individual's outcomes, with the added benefit that an adolescent, unlike an eighteen-month-old, is linguistically and socially competent to be an active participant. 'We need to challenge the widespread assumption that . . . interventions early in life are more worthwhile and more cost-effective than later ones,' Sarah-Jayne has said. 'Of course early interventions are vitally important, but if a child has "slipped through the net", it's not too late to provide extra support in adolescence.'

Immersing myself in Sarah-Jayne's award-winning book, *Inventing Ourselves: the Secret Life of the Teenage Brain,* left me with the conviction that, as parents, the most useful thing we can do to help our teenagers navigate the turbulent changes prescribed in their biology that are taking place in their brains and bodies is to act as good role models, quashing our own bad habits where we can while trying to calmly observe our own children's destiny unfold. Beyond that, it might be beneficial to try to facilitate their exposure to new safe experiences and activities so that they can experiment and discover what they like to do, surrounded by peers who are also positive and active. But being the parent of a teenager seems to require accepting that they are shaking off family influence in pursuit of authentic selfhood. So, really, I'm just hoping Sarah-Jayne's wise words will see me through those potentially turbulent years with my sanity and sense of humour intact.

All downhill from here?

By the time an individual is in their late twenties or early thirties, the prefrontal cortex has long been fully integrated into the connectome. The twin powerhouses of synaptogenesis (the generation of synapses that takes place at such a rapid pace during the child's early years) and synaptic pruning are calming down. During the next decade many of us are at our physical peak, in terms of both brain and body. This period can be among the busiest and most fruitful of our lives, with experiences rapidly logged in our social, sexual and intellectual lives. Past this point, it all starts irrevocably to decline, or so some would have us believe.

But in neurobiology, as in life, there are compensations to

ageing. Low-level cognitive functions, such as reaction times, and many higher-level ones, such as on-the-spot reasoning, may be slower for those of us who have passed our thirty-fifth birthday, but another cognitive function keeps improving throughout our lives. 'Crystallised ability' includes things like breadth of vocabulary and knowledge of the world, and it just gets better as we age. Our brains are wired to create wisdom, to accumulate a bank of experience and memory that adults, and especially older adults, can draw upon.

I wondered, though, how the neuroscience of wisdom fits with a contrary view of some older people's behaviours, which can make them appear rigid, stubborn or narrow-minded. There must also be a neural basis for a reluctance or inability to take on new ideas. And what, I wondered, determined which older-brain profile we were likely to assume? Was I destined for wisdom or Victor Meldrew-style crotchetiness?

I mounted my bike to visit another Cambridge don and pedalled upstream along the river to meet Dr Rogier Kievit at his office in the Medical Research Council's Cognition and Brain Sciences Unit. Rogier is best described as a quintessential Dutch man, exceptionally smiley, youthful and healthy-looking. He specialises in the ageing brain, and I wanted to ask him more about the neural basis for wisdom and rigidity in the older brain profile.

I've known Rogier and his wife Anne-Laura, who is also an award-winning neuroscientist, for some time. They have a child close in age to mine. We see each other often, but though we talk about work in passing it was the first time I had asked Rogier in detail about the implications of his research. He lost no time in teasing me: I was evidently thinking, slightly anxiously, he said, about the fate of my own declining brain. Then he pointed out that rather than positioning wisdom and rigid thinking as opposites, you could conceive of them as being essentially the same thing. If you reframe 'rigidity' as 'expertise', you can see how an older brain's preference for sticking with its tried and trusted cognitive strategies might be a winning strategy. An older person will have acquired dozens of areas of expertise over the course of their life and, taken together, they could be said to add up to wisdom.

'Say you're a professional tennis player,' he said. 'You learn to hit the ball in a specific way and you win, so you continue. We could describe this tennis player as being stuck in their ways or we could say that by accumulating expertise they have finessed their skill to the point where they can play at an amazingly high level with almost no effort. It might be the case that older people find it difficult to adapt but that is the flip side to the accumulation of wisdom.'

So, wisdom is the pay-off for lifelong learning and typically balances out older adults being less motivated than those thrill-seeking teenagers to seek out novel experiences and information. They just don't need them as badly. But even if we no longer want to explore new countries or take up the bass guitar, we all need the basic neuroanatomical capacity to acquire and store new information until the end of our lives. The precise molecular process by which the brain manages to do this is only now being understood.

Rogier's work involves observing changes in the brains of adult volunteers as they carry out learning tasks. He told me that the brain actually bulks up with training, much as muscle does. All those arborisations of the neurons take up physical space. But it obviously can't be the case that your brain expands every time you learn something new, since there is only so much space within your skull. So what is happening? Rogier told me that it's only during the initial learning process that you get brain expansion. Once you've acquired the skill, circuits are refined to their key pathways, reducing the brain's volume once again. This is plasticity in action, with the brain sculpting itself so that it can increase neural capacity with maximum efficiency.

Though this phenomenon has been observed, it's only very recently that the precise mechanism by which the brain accomplishes it has been pinpointed. Mriganka Sur, of the Massachusetts Institute of Technology, established that, once a connection between neurons has been strengthened to a certain point, a genetic switch is triggered that causes neighbouring connections to dissolve. In this way the brain optimises its circuits and settles into efficiency. As a brain ages, it relies more and more on these tried and tested pathways.

When it comes to the all-important business of processing new incoming information to help produce our unique view of the world and our place in it, the older brain also behaves differently from the younger one. It assigns less importance to the new incoming signals it receives, via the ears, eyes and other sensory organs, in comparison with its prior experience and expectations. Again, this strategy makes sense: those systems for collecting information from the outside world will at some point start to fail. The brain has already spent vast cognitive energies building up its stock of experiences and storing the memories, testing mental strategies and finessing them. The older brain acts efficiently, placing more value on past experience and knowledge than new.

Talking to Rogier had reassured me about what I had previously regarded as little more than general decline. There seem to be upsides to the process in the form of wisdom and expertise, and I felt less concerned that an old age of grumpy narrow-mindedness was all but inevitable. For one thing, as Rogier was at pains to stress, there is huge variation among individuals when it comes to developing the more bothersome effects of an ageing brain, such as memory loss or woolly thinking.

Neuroscience is discovering more and more about the effects of ageing, and its understanding of dementia is already well advanced.

Dementia, like obesity, is one of the scourges of contemporary life. It is usually progressive, becoming severe and eventually terminal, and it's estimated that 7.7 million people will develop the condition around the world every year. It can be devastating to cope with as it snuffs out memories, along with the ability to live an independent life, eroding the sufferer's personality and history. There is a genetic component in a small proportion of cases and the NHS offers genetic screening for some high-risk families, but studies have also implicated lifestyle issues, including obesity, low physical-activity levels, depression, lack of social contact, smoking and leaving education early as contributing risk factors. Either way, the underlying mechanism causing the debilitating symptoms of dementia is nerve cells dying off, be it through abnormal clumping of proteins that cause tangles in neuron processing and blockages in cell bodies, or restricted blood circulation in the brain.

Until the late 1990s it was believed that the brain had virtually no weapons in the face of such an onslaught against its neurons. But then an incredible discovery was made. Researchers at the Salk University, California, led by Professor Rusty Gage, revealed that physical exercise induces the birth of new cells in the brain. Conventional wisdom at the time said that a person is born with all the neurons they'll ever get, meaning that dementia, or indeed any debilitating brain condition, was effectively a slow death sentence, impossible to evade. But actually neural stem cells, small precursors to nerve cells, exist even in the adult brain. They can be found in the hippocampal region, buried deep at the centre of the brain: an area fundamental to learning and memory. In a wonderful set of experiments performed initially in mice, it was demonstrated that movement induced these stem cells to develop into fully formed neurons through a process called neurogenesis. Even more incredibly, the simple act of combining physical activity with exploring a new environment and interacting with different individuals helped these newly born neurons to integrate fully into the existing circuit, to survive and flourish. Essentially movement provided the mechanism for forging new neural networks and new ways of thinking.

The implications of this unprecedented discovery were immediately seized upon, with further study indicating that the findings might translate to humans. There is now a robust body of evidence to suggest that this is the case, though the results of one very recent study question the extent to which neurogenesis continues to occur in our species into adulthood. More research will refine the circumstances in which it can take place but, for now, I find it comforting to believe that staying physically active might provide me with the power to rejuvenate my brain. Exercise certainly helps to protect existing networks by decreasing levels of the stress hormone cortisol that can lead to cell-connection death if it stays at high levels for a prolonged time. It also increases the production of certain brain chemicals, including endorphins, dopamine and serotonin – critical neurotransmitters associated with feelings of pleasure, reward, motivation and improved mental wellbeing. Basically, exercise works as a natural antidepressant, which helps to explain why virtually all ageing neuroscientists, including me, go running.

There has recently been more good news for our prospect of maintaining healthy brain functioning into old age. During the course of my conversation with Rogier, he highlighted a paper by researchers from seven institutions in Berlin led by Dr Ulman Lindenberger, managing director at the Max Planck Institute for Human Development. The study compared the overall health of more than 300 elderly people tested 20 years apart, half in the period 1990–1993, the others in 2013–2014. Although the individuals analysed were matched on every measure the researchers could think of (age, gender, education, where they lived, physical health) the difference was striking: older people now have much better memories, are happier and have better morale. Rogier's interpretation was that the slow but steady improvement of education and public health seems to have led to happier, healthier older people, which shows that our rate of cognitive change in old age isn't set in stone. 'In other words,' as Rogier emphasised, 'given that there have been no meaningful genetic changes between these generations, this huge improvement is likely to be entirely due to environmental factors.'

Before I left the Medical Research Council building I asked Rogier, who has dedicated his career to investigating the natural ageing process in the general population, what he did to protect his own brain from its effects. We compiled a list, based on his thoughts, of things we can all do to build our brain's resilience as we age. I can't help relishing the delicious irony of setting out to thwart our fate by empowering ourselves with more knowledge derived from neuroscience. Top of the protection tips is, surprise, surprise . . .

1. **Being physically active.** It doesn't have to be running. Thirty minutes of a low-level workout such as walking, swimming or cycling three times a week is great for the brain and body. Whatever your size, whatever your timetable, get out there and be physically active. It will not only

potentially ramp up your neurogenesis but also keep your brain capillaries healthy.

2. **Get a good night's sleep.** There is mounting evidence to show that sleep helps to consolidate connections between neurons, enabling a host of processes, such as turning new knowledge into banked memories. Sleep also gives your immune system the chance to clear away any toxins made in your brain during the day so that they are less likely to accumulate and kill off neurons.

3. **Stay socially active.** Spending time with friends and family, discussing things, learning from other people, taking on board their perspectives and ideas helps to keep your brain processes dynamic and is generally associated with better wellbeing.

4. **Check your diet.** Any food that is associated with poor cardiovascular health (animal fats, processed foods, too much sugar) is also associated with poor cognitive health. The general rule is to eat for your heart and brain as one. This protects against micro-strokes that could asphyxiate neurons.

5. **Keep learning.** Learning early on in life helps to protect against cognitive decline later in life. Research shows that the longer people stay in education the more likely their brain will age more healthily. But lifelong learning of any kind, inside or outside formal education, is a great strategy for maintaining brain health.

6. **Stay positive.** Believing that you have a poor memory is associated with a swifter decline in performance. If you start

to avoid new social settings, for example, because you're worried you won't remember names or know how to navigate spaces, this can accelerate the downward trajectory. And, generally, better mental health is associated with better cognitive health: if you're feeling blue you're less likely to find the motivation for, or derive pleasure from, exercise, looking after yourself or getting out and about for social interaction. Writing a gratitude journal each night before you go to sleep makes it easier to wake up in a more motivated mood, keen to replicate some of the adventures you experienced the day before and seek out new ones.

I ended my afternoon of conversation with Rogier in a different frame of mind from the one in which I'd arrived. The neuroscience of ageing, far from being a picture of doom and gloom, lends itself to positivity. It's an area in which revolutionary breakthroughs such as the discovery of neurogenesis and neural-pathway refinement have stimulated decades of well-funded research into the development of new treatments. It's also an area in which measures we can all take as individuals have demonstrable beneficial effects on our own personal outcomes.

Throughout this chapter we have been examining the sorts of changes we can expect that our brain will undergo during our lifespan and have touched on the individual's stability of personality traits and temperament. There will be many more breakthroughs in the near future as more rigorous individual predictions become possible. There is a massive project now under way, for example, utilising technologies that make it possible non-invasively to scan the brains of babies in the womb with relatively high resolution: scientists can see the resting-state pattern of brain circuit connectivity in the baby even before birth, and have established that its connectome is more similar to its mother's, on average, than to other people's. It looks as if we will need to extend our understanding of inherited qualities from individual genes to their impact on the map of our brain circuits. Scientists will soon be able to follow an individual from before they are born through to death, to map their brain's circuits as they develop and try to tally it with behaviour and life trajectories.

Later we'll be exploring this project and its implications in more depth, examining how biomarkers, genetic tests, brain scans and EEG readings can help to make more robust predictions about an individual's outcomes. We may soon be able to identify those individuals most likely to act in a dangerously impulsive way during adolescence, or those who have a greater propensity to develop addictive habits. We will know more about the factors implicated in resilience in the face of life's challenges and understand why some people live well into their centenary years with the mental agility and focus of a mountain goat . . . But for now, armed with our greater understanding of the neurobiology of life's broad stages, it's time to look at how one of our basic behaviours – eating – is generated in the brain. Though 'basic' never means simple when it comes to brains, and there's certainly nothing simple about how we choose what to eat.

CHAPTER THREE

The Hungry Brain

Everybody eats, but *what* we eat is deeply personal. Our food choices are linked to our emotions and to our sense of identity, to our aspiration to be healthy and the sometimes conflicting desire to indulge. Food has become a complicated subject for many of us. It can be a source of great pleasure but also much anxiety. What can neuroscience tell us about how we behave around food? To what extent are we able to choose, freely and consciously, what we put into our mouths?

In this chapter we'll be building on what we're discovering about how our brains shape our behaviour to look at how we decide what to eat. We'll be unpicking the interplay between innate preference and free choice when it comes to, say, kale versus doughnuts, and disrupting the idea that it's all a matter of what you happen to like or, indeed, what you can force yourself to resist. Even this most universal of behaviours is also fascinatingly complex and, I would suggest, much less a matter of free choice than we tend to assume. We will examine how human beings evolved to find certain foods compellingly delicious. Our need to eat and drink is, of course, fundamental to us but it is also motivated by desire. Our brains are hungry for more, driven by our instinct to seek out activities and tastes that we find rewarding. So we will also take a closer look at the specifics of the reward system to understand how we derive pleasure from gratifying our appetites.

There is huge variation in behaviour around food choices, which can serve to convince us that what we choose to eat is highly personal. It's easy to assume that it's all a question of taste. This is true, of course, but by no means the whole picture. Human beings generally love salty, fatty, sugary foods, the high-calorie treats that many of us find virtually irresistible, even though we're aware that too much of them can be a bad thing. At the population level, we're hard-wired to go for that doughnut over the kale. Even as an individual, your preferences are not merely a matter of what you've learned to like through exposure to the foods around you. Your food choices, and your willpower when it comes to resisting temptation, are massively subject to external influence, a fact that food manufacturers and retailers are adept at exploiting. If you have ever fallen prey to the supermarkets' trick of piping the smell of baking bread to the store's entrance to entice you in, you may well have struggled to resist their wiles and control your appetite as you shop.

A host of mechanisms and influences contributes to your decision-making around food, many of them much less obvious than that one. Some are buried deep in the well of what John Bargh, a professor of psychology at Yale University and pre-eminent expert on the unconscious mind, called 'the hidden past'. You might be surprised to know, for example, that what your paternal grandfather liked to eat, decades before you were born, has an effect on your food choices today.

A lot of research is going on in this area – unsurprisingly, given that the obesity epidemic is the most urgent public-health crisis of our times. Experts warn that if trends continue, by 2025 roughly a fifth of the world's population will be clinically obese.

I won't spell out all the reasons why that's a big problem. Let's sum them up by saying that a diagnosis of clinical obesity reduces lifespan by ten years, on average.

We are typically led to believe that our food choices are conscious, and that if we cannot take 'good' decisions about what to eat we are somehow to blame. Attitudes towards obese people have become increasingly judgemental. They are seen as lazy, greedy and lacking in willpower. It's a simple maths exercise, right? They should just eat less and move more.

Neuroscience has a lot to teach us about why this view is too simplistic. Recent advances in brain-imaging technology have yielded breakthroughs in what we know about how appetite is shaped and controlled in the brain. In the past, neuroscience made most of its leaps in understanding by studying anomalous behaviour, deviations from 'normal' brains brought about by illness, such as a stroke or a catastrophic brain injury. Now healthy brains in living mammals can be studied as they carry out their routine business. This means that we can start to get to grips with what is happening in my brain and my 60kg body to make me order a delicious flaky pie rather than a salad for lunch.

The results seem to suggest that, bottom line, our appetite, at both species and individual level, is largely determined at birth, written into our genes and pre-wired into our brain circuitry. It is shaped by biological traits that have evolved over millennia to drive us to find certain foods delicious.

But it's not quite that simple. Obviously I don't order pie every day. In general my food preferences are fairly static, but I also enjoy variety on my plate. There is also scope for personalisation in our food preferences: some people walk straight past the savoury options, not even tempted, but are hopelessly drawn to the cake stand. The point is, it may seem obvious that *you* find certain foods more appealing than others but when you start to understand why this is true, the ramifications for us as individuals, and for society, are vast.

Kale or doughnuts: how do we choose?

To understand how and why we choose what we eat, or indeed why we do anything at all, we need to know more about how thoughts and decisions arise in our brain: in essence, how consciousness is generated. Experiments conducted almost a century ago deduced that it derives from the constant buzz of rapid electrical pulses racing along neurons, using chemical neuro-transmitters to bridge the synapses and activate the next neuron in line. In essence our brain, like the central nervous system of any species on Earth, is simply an electrochemical circuit board. It's considerably more complicated than that makes it sound, though, given its scale.

Until recently scientists had very little idea how this immensely intricate structure achieved the feat of neatly converging billions of travelling electrical signals to produce functions such as decision-making, emotion or memory, but technological advances have now enabled scientists to observe these staggeringly complex processes in live organisms.

Genetic engineering is a key technique that has been funda-mental to these advances. It has allowed us, for example, to tag the individual building blocks of the brain – the neurons. Increasingly sophisticated microscopy (another crucial advance) allows us to visualise them at high resolution. By combining these techniques it is now possible to map the brain circuits that are implicated in specific behaviours.

Studies are typically carried out in simpler organisms, such as the fruit fly, worm and mouse, but they are still immensely informative about our own minds, thanks to the surprising degree of similarity of brain architecture and operating system across species. All brains, however simple or complex, use neurons as their building blocks and all animal species on Earth use pretty much the same system of electrochemical communication. (Which is why you can take a faulty gene from a person with Parkinson's disease, express it in a mouse, and the mouse will develop tremors reminiscent of the disease in humans.) A mouse and a human are obviously not the same but it is possible to extrapolate the results from simpler organisms to humans and combine them with other findings from different fields to understand how behaviours arise and how we can help individuals and societies to flourish.

Genetic engineering made possible the invention of Brainbow Mouse, a breakthrough that is yielding massive insight into how behaviour is generated at a neural level. Brainbow Mouse, a genetically engineered organism created in 2007 by Jeff Lichtman, professor of cell and molecular biology at Harvard University, received his catchy name because his brain is illuminated with all the different colours of the rainbow. (I say 'his' but really I mean 'their' as there have been hundreds of generations of Brainbow Mice born since 2007.)

A specific circuit in the original mouse's brain was modified with genes extracted from a species of jellyfish that glows fluorescent green as a warning to predators. The gene coding for Green Fluorescent Protein (GFP) was isolated, tweaked so that it was expressed constantly, cloned, then introduced into a distinct area of Brainbow Mouse's brain. This tagging of individual neurons allows us to view the cells in isolation and observe the complexity of their connectivity across the brain. It was suddenly possible to map brain anatomy and function in an unprecedented way. The results have helped inform the new field of connectomics: mapping the connectome, the pathways of thought. Lichtman has essentially created a wiring diagram for a conscious brain in action.

This technique, alongside others such as optogenetics (which we'll discuss later), has deepened our understanding of how specific systems in the brain supply us with, for example, feelings of motivation and reward. So, what exactly does the reward system, or hedonic pathway, do? How did it evolve and how does it bring about our decisions on what to eat?

Scientists have been studying this system for more than sixty years, since it was first accidentally discovered in 1954: researchers found that a rat will press a lever hundreds or even thousands of times an hour to obtain stimulation to this circuit, stopping only when it's exhausted. Humans, it transpired, will behave similarly. This is because the reward system, in evolutionary terms, is an ancient piece of kit, conserved across species. So, in a rat, mouse, dog or cat it is structurally and functionally pretty much the same as yours or mine. It developed to facilitate our survival, to keep us motivated to expend the precious energy needed to stay alive and reproduce.

It consists of three main pathways. First, there's a tiny cluster of nerve cells buried deep in the midbrain in an area called the ventral tegmental area, or VTA. This is where the chemical dopamine is produced, which travels up to another area of the brain called the nucleus accumbens, a peanut-shaped structure that lights up with electrical activity in response to dopamine. This circuit sparks to life whenever we experience pleasure. Merely thinking about the activities we derive pleasure from, including eating and having sex, is enough to activate it. Cleverly, the circuit is also sensitive to exercise, helping to motivate us either to hunt out more food, have more sex, or run from predators. This brain region basically facilitates all three of our essential life goals. (Well done, evolution; very efficient.)

The nucleus accumbens and the prefrontal cortex (the area of brain directly behind your forehead, involved in higher executive functions, including reasoning, planning, flexibility of thinking and decision-making) are connected. This ensures that we remember feelings of pleasure and associate them with the right triggers, thereby motivating us to repeat the experiences.

Interestingly, drugs of abuse hijack this system, which is why they can be so addictive. (Not so good, evolution: an unfortunate side effect there.) And though it's not true to say that sugar works like heroin, say, or alcohol on the reward system, as is sometimes claimed, it is true that there are flaws in the system of checks that exist to put the brakes on insatiable appetites. To control for satiety, the stomach passes a signal to the brain instructing us to stop eating because it is physically full. The problem is, the braking system isn't sufficiently responsive. Its effects are frequently felt too slowly. Bariatric surgery, in which a band is fitted to shrink the stomach, is a last-ditch attempt to boost the sensation of satiety and help the person restrict their food intake. As a species we are just not very good at deciding when we've had enough of anything we find delicious. More is always more, as far as our bodies and brains are concerned, and it's the reward system that drives this behaviour.

The reason for this is that our reward systems evolved for an environment vastly different from the one we have created for ourselves. Mammals basically evolved, over roughly 250 million years, to carry on eating no matter what. Anything that enhanced our individual ability to seek out food, consume it quickly, carry on eating even when we were full, store fat more efficiently or hold on to our fat stores for as long as possible gave us an advantage. These traits thrived – they were more successfully passed on to offspring through our genes. At the same time it was a good idea to be lazy, except in very specific circumstances: we evolved with the motivation to expend energy to seek out food, eat it, then reproduce, but that's about it.

Most of us in the developed world are surrounded by the potential to eat all the time. We don't even have to leave our houses to find food. At the click of an online order button it can be brought directly to us. There's really no need to eat as if we might not get another meal for days but the biology underpinning our behaviour with food and instructing us to keep eating is still there.

Is humanity hard-wired to overeat?

With the development of genetic-engineering techniques, current research into the science of appetite is subtler and more potentially applicable in a therapeutic context than ever before. I wanted to dig deeper into what it had to say about the part played by inherited traits and hard-wired biological structures versus environmental factors, so I took to my bike, on an exceptionally blustery and wet fen day, and cycled over to visit Dr Giles Yeo, who has been studying the genetics of obesity at his laboratory in the Cambridge University Metabolic Sciences Unit at the city's hospital for nearly twenty years.

I first met Giles more than a decade ago and he hasn't changed much. He's still sprightly, energetic and enthusiastic. He suggested we talk over lunch, so we could combine our conversation on appetite with the intake of calories to power our brain activity during the discussion.

We walked down to the hospital canteen and I'll admit to being acutely aware of this obesity don's presence as I selected my food. I opted for a healthy salad and fruit. We passed his laboratory, where gene-amplification machines whirred, copying pieces of DNA by the thousand, and pipettes busily sucked up homogenated genetic samples. Eventually we arrived at his rather chaotic shared office to get stuck into lunch and conversation.

Primarily I wanted to know what Giles thought about personal autonomy when it came to food choices, in the light of his research into the links between genetic inheritance and obesity. 'The question of free will is,' he explained, between mouthfuls of his avocado and prawn sandwich, 'one of intense interest to me. Does our biology provide us with an excuse to be lazy and overeat? Unfortunately, the answer is, at least for many of us, yes.'

As Giles put it, if you strip us down to the essentials, we have three primary biological drives in our lives:

1. Find and eat food
2. Avoid becoming food
3. Reproduce so that all this can continue

'Basically,' he said, 'there's a fundamental drive that has evolved to help us to achieve these goals. It's pretty hard to ignore.'

I pushed him on this. Was he saying that we're all, to a greater or slightly lesser degree, designed to overeat? If that's so, why isn't everyone fat? Why does the modern 'disease' of fatness seem to affect about half of us? Why are some of us more vulnerable than others to the predicament in which we find ourselves?

The answer, essentially, is that on a species level we're all in the same boat but at an individual level there's still room for a heck of a lot of variation. There are close to 150 genes implicated in predisposing your weight and body shape. They include those that direct how hungry you feel (shaping the sensitivity of receptors in the stomach, sending signals to the brain instructing when to eat or when satiety has been reached); the genes involved in the hedonic circuitry (some people simply require more calories to tickle the reward pathways in their brains: their receptors have less sensitivity so they have to eat more – two pies instead of one, say, for that burst of pleasure); and the genes involved in how our brain senses the levels of essential nutrients in our bodies, instructing us to keep on eating if we have too little.

In the past there was little genetic pressure to stop people from becoming obese. Genetic mutations that drove people to consume fewer calories were much less likely to be passed on, because in an environment where food was scarcer and its hunting or gathering required considerable energy outlay, an individual with that mutation would probably die before they had a chance to reproduce. Mutations that in our environment of abundant food now drive us towards obesity, on the other hand, were incorporated into the population. Things are of course very different now but the problem is that evolutionary timescales are long. It's only in the last century or so, approximately 0.00004 per cent of mammalian evolutionary time, that we managed to tweak our environment to such a degree that we can pretty much eat whatever we want, whenever we want it. Evolution has another couple of thousand years to go before it can catch up with the current reality of Tesco Direct and Deliveroo.

But what if we could intervene genetically to place a check on our ravenous reward system? That's what Giles is currently researching. 'In the past there was no real evolutionary benefit to having a less sensitive reward pathway but, of course, now there definitely is. The question is, could we and should we step in to engineer it?' Could you, for example, genetically tweak someone to dampen down the pleasure they get from sugar, so that they no longer feel the urge to add it to their tea? Could you strike a balance between ensuring that an individual doesn't lose their small pleasures even as you improved their health prospects and increased their genetic fitness? In essence is there any imminent possibility of engineering ourselves in real time so that we can outrun the hazards of our environment?

For adults living now, the short answer is no. But for those born in the future, it is a potentially plausible scenario. CRISPR/Cas gene editing is a revolutionary technology that enables genes to be manipulated more cheaply and easily than ever before. If you wanted to engineer a genetic change across an adult organism – to remove predisposition to obesity, say – you would have to engineer every cell in the body, a task that's impractical. But CRISPR/Cas editing can be applied to human embryos, where the scale of the task is much smaller. Indeed, CRISPR/Cas is being used in labs around the world by researchers investigating the eradication of specific genetic conditions, such as B-thalassemia, an inherited blood disorder. To apply the technique to the issue of obesity, though, even with embryos, would require genetic engineering on a vast scale due to the large number of genes involved, and would be beyond the scope of what is currently permitted by regulatory bodies. As Giles put it, 'It would necessitate the engineering of embryos, and discarding of embryos, on a massive scale. I think we're a long way away from accepting that as a society.'

In any case, the technique is still in its infancy and has yet to be proved safe long-term. We will be returning to the practical and ethical conundrums it poses later in the book when we assess the ways neuroscience is being applied in the real world, and consider the case of the Chinese scientist who in autumn 2018 announced that he had applied the CRISPR technique in twin baby girls conceived by IVF. But for now let's return to the issue of obesity

Isn't there one stand-out gene implicated in obesity that could be edited, that might make a difference on a shorter timescale, I asked Giles? 'Well,' he said, 'there's FTO, short for Fat Mass and Obesity-associated protein.'

It transpires that half the world's population have the version of this gene that makes them 25 per cent more likely to become obese. If somebody has two versions of this genetic variant of FTO (which a sixth of the world's population does) they are probably three kilograms heavier than they should be and have a 50 per cent greater risk of becoming obese. This gene is expressed not in the circuits that make up the reward system but in the hypothalamus. It feeds into the reward system, though, by instructing the body that more nutrients are required. It works to keep the person awake and eating. Essentially, there's a system in place that explains why so many of us find ourselves raiding the fridge late at night, rather than going to bed. But we are still a way off engineering our way out of the problem via the FTO variant. Many of us would in any case shy away from the prospect of genetically hacking ourselves, or our children, in order to circumnavigate our propensity to be obese. More research is required to turn this promising discovery into a new way to treat the obesity epidemic afflicting our species.

If technological leaps via genetic engineering are not yet a viable option, is there anything we as individuals can do? Adopt a rigorous exercise programme, maybe? Unfortunately, exercise alone does not usually mitigate for having a genetic predisposition to becoming obese because a person's metabolism may also be predestined to be slow. Giles confirmed it. 'For people with this double FTO genetic variation, even if they're active it can be almost impossible to keep a healthy BMI.'

I left Giles's lab having got the confirmation I'd been expecting but it was pretty uncomfortable. Our individual appetites are largely programmed by circuitry that's arisen out of millennia of evolution to bequeath us a unique genetic package, but in general, our brains have evolved to drive us to seek out high-fat and high-sugar foods. How strong this urge is for any individual depends on the genes and brain wiring they were born with. A person's attempts to change their eating habits are always going to be constrained by these factors, which explains why weight loss can be such a challenge for so many of us.

Healthy eating starts in the womb

Our behaviour around food is not solely shaped by our genetics. Recent studies suggest that, in terms of our body weight, 70 per cent is estimated to be directly shaped by the genes we were dealt. But that leaves a whopping 30 per cent accounted for by environmental factors. It transpires that it's possible either to tweak these deep-brain circuits or to reinforce them during our very early years by altering our environment. During the forty weeks of gestation, the foundations for a baby's brain, including the reward system and all the other regions involved in control of appetite, are laid down in a highly dynamic process driven by the mother's and father's genetic direction. But in addition to genes, the in-utero environment in which this happens can also shape the developing brain.

Professor of biopsychology Marion Hetherington, who works at the Human Appetite Research Unit at the University of Leeds, has been analysing how a mother's diet can shape her baby's food preferences and appetite in later life. When I spoke to her she highlighted research from her lab and others around the world that tells a story of windows of opportunity, when it is possible to direct an individual away from their predestined pathway to obesity.

Most of us, especially if we have been pregnant, are familiar with the idea that what a woman consumes during pregnancy is important for her unborn child's wellbeing. Pregnant women are told to limit caffeine, consume no or very little alcohol and completely cut out nicotine, any drugs of abuse and foods potentially harbouring dangerous microbes, such as unpasteurised milk and cheese. Elements of the maternal diet can be transmitted through the amniotic fluid surrounding the foetus in the womb, and later through the breast milk, to affect the baby's rapidly developing brain. Experiments have shown that if a pregnant mother eats foods rich in highly volatile compounds, such as garlic and chilli, her newborn baby will orient itself towards these familiar smells and flavours, moving their heads and mouths towards the source. It is not yet known exactly how previous exposure to these tastes affects the baby's brain circuitry but it seems reasonable to conclude that our good friend the reward system is at the centre of things, as usual. The baby's brain has learned to associate specific smells and flavours with the rewarding comfort of their mother.

You can see the same effect throughout early-years development. If breastfeeding mothers continue to eat particular foods (one study focused on caraway seeds) this information will be transmitted through their milk. Even years later, their children will have a preference for these tastes, opting for hummus with caraway seeds rather than plain hummus. These studies have been replicated using different experimental paradigms and, taken together, they strongly indicate that if a pregnant and breast-feeding mother eats a healthy, varied diet, the exposure primes the baby, imprinting them with a preference for a good varied diet themselves later in life, potentially all the way into adulthood.

Weaning represents another opportunity to shape food pref-erences. As the baby develops into an infant and solid foods are introduced, it appears to be possible to prime them to enjoy eating vegetables, over rice cereal or potato, by adding vegetable purée to expressed breast milk. Babies exposed to carrots and green beans will smile when offered it again, and eat more of it.

Thinking of my son and whether I'd done enough to instil in him a preference for salad over chips, I asked Marion whether there were further opportunities later in childhood, or whether the window closed once weaning was complete.

She smiled as if it wasn't the first time she'd had this conversation with a slightly anxious parent. The basic rule of thumb is the earlier the better, she told me, but there is room to make a difference throughout a child's early years, even up to the age of eight or nine. 'The key is to be persistent and positive. You need to offer a novel food, such as a new vegetable, for example, eight to ten times before the baby or child will start to associate pleasure with that specific taste. But, yes, it is possible to tap into our innate reward systems and use them to our advantage.' Older children can be helped to like broccoli, for example, through association cues, so that they link eating nutritious little brassica florets with treats such as going to play in the park or receiving lots of praise or stickers.

But surely there's a problem with viewing childhood as a chance to change individual genetic destiny. Isn't it likely that those parents with a genetic predilection for more processed food instead of vegetables will find it harder to eat a good varied diet during pregnancy, breastfeeding and weaning? If I don't like broccoli, I'm not very likely, as a sleep-deprived new parent, to go to the unappreciated effort of buying, preparing, cooking and serving broccoli to my child if nine times out of ten it will be tossed onto the floor or left uneaten. In the real world as opposed to the lab, is it not more likely that an individual's inherited hard-wiring around appetite would be reinforced rather than modified by early environmental factors?

'That's true,' Marion admitted. 'It's often a missed opportunity. If you're dealt genes that incline you to obesity and are then surrounded by an obesogenic environment, continually offered unhealthy foods by your parents, and as a family live a sedentary lifestyle, then the path you're veering towards is inevitably one of obesity.'

Marion is trying to tackle this head on. She's been working with manufacturers of baby food to develop more vegetable-based products and market them as ideal foods for babies who are just starting on solids. Not every parent will reach for them, but some will.

So, it seems that if you're a parent you may be able to make useful improvements to your children's outcomes (so long as you can avoid the sinking feeling of yet another thing to feel guilty about). But what about us adults, who are obviously well past the early years? Is there any way for us to rewire our brains, so that we choose healthy over unhealthy options? How does the brain's famous plasticity affect our ability to alter habits around food choice? Even if it gets harder to rewrite everything that's been reinforced by behaviour over the years, it can't be impossible, surely. Some people do lose weight and keep it off, or become vegetarian or vegan.

The research confirms Marion's summing up of the situation: though it's never too late to change our behaviour, it gets harder as our habits become more entrenched, and relying on willpower alone probably won't be enough for most of us. For a start, will-power is not some fixed moral quality to which we all have equal access. An individual's capacity for self-denial is as much the product of innate neurobiological and environmental factors as any other personality trait, and fluctuates according to dozens of contextual conditions – if you're tired you have less willpower than when you're well rested, for example. 'White knuckling' – as Alcoholics Anonymous term it when an alcoholic is dependent on willpower and stuck in endless moment-to-moment resistance to drinking – is not a sustainable strategy for any habit change.

Weightwatchers, with its group support and accountability, is frequently cited as the most effective strategy for sustainable weight loss. Their programme uses techniques that have been proven to increase people's chances of sticking to a diet, such as surrounding yourself with healthy and positive friends, exercising as part of a social group to keep up morale and planning treats for yourself as you hit milestones in your healthy-eating programme. 'Eat Right Now' began as a mindful eating programme developed by Dr Judson Brewer, an addiction specialist first at Yale University and then the University of Massachusetts. It has proved highly beneficial for its participants, reducing their food cravings by up to 40 per cent, and is now being offered by other university health programmes.

Different strategies work for different people, because habit formation is complex and varies from person to person. This is not surprising since it arises from a complex interplay between three factors: the ancient brain circuitry we evolved with as a species, the genes we were born with as an individual and the environment we currently find ourselves in. As a result, if we want to modify our eating behaviour, it helps to commit to experimenting on ourselves until we find the things that work for us. There's never any one-size-fits-all solution.

Evolution, epigenetics and eating habits

My conversation with Marion had reinforced my sense that there is at least a bit of wiggle room to create change in our own behaviour around food. I knew that the emerging field of epigenetics was causing a lot of excitement among researchers into appetite. How close had they come to developing a new therapeutic route by which adult appetites could be shaped in later life? To find out more about epigenetics and its potential applications, I visited Cambridge University's Pathology Department to see Professor Nabeel Affara, who has been studying how our environment triggers alterations not in the DNA code itself but in how it is read and used in the body – gene expression. The exciting thing about this is that genetic mutation can be observed within a couple of generations rather than over the long timescale of evolutionary change.

The role of environmental factors in directing gene expression has only recently been discovered and is called epigenetic regulation. Epigenetics helps to explain how the cells in your body can behave in radically different ways despite having the same genetic code. Each cell in your body has machinery to translate your genetic code into the proteins it needs to do its specific job. These 'volume dials' for the DNA alter depending on the environment: the stomach environment instructs a cell to act in a particular way, and your eye socket tells the cells there to behave like eye cells.

I met Nabeel at his departmental reception, the air thick with the acrid smell of burned agar plates. Nabeel is investigating how your parents' (or even your grandparents') diet can affect your behaviour and that of your children. He studies the stage prior to conception, looking at how the nutritional environment of the sperm and egg can alter the way in which genes are expressed in the next two generations.

The epigenetics of appetite has been shaped by long-term studies of Dutch populations born as the Second World War was ending. They have compared the health outcomes for children born to families under German occupation, where people virtually starved for a year between 1944 and 1945, with those born in the unoccupied areas who had far greater access to food. It transpires that being born to parents whose nutritional status was seriously compromised at their child's conception means that the child is far more likely to develop obesity and diabetes later in life. The explanation lies in the mismatch hypothesis. If a baby develops in an environment of scarcity, their metabolism struggles to adjust to conditions of plenty. It's not that their DNA code has been modified by those conditions, however harsh, but the way their genes *behave* has been altered, and that modification is passed on to the next generation, and the next. This effect is worth considering as we survey our current calorie-rich but sometimes nutrient-poor environment.

Is this yet more evidence for the hard-wiring of our appetites, not just before birth but even before conception? Well, yes. But other epigenetic research, while still at an early stage, might eventually lead to treatments from which an individual adult could benefit. There is mounting evidence that all sorts of behaviours are shaped by the environment in which our parents lived before we were conceived. One particular experiment produced results that may be transformative in the treatment of addiction. In fact, their implications are so far-reaching that they left the whole scientific community reeling when they were published.

Kerry Ressler, professor of psychiatry and behavioural sciences at Emory University, has been looking at how environmental pressure can change behaviour around food choice in mice. As we've already seen, mice and humans have similar reward pathways, with the nucleus accumbens lighting up in anticipation of a sweet or fat-laden reward. Nearby brain regions, the amygdala and insula, are associated with emotions and, in particular, fear. Kerry's studies manipulated the interplay between these basic systems. The mice were delivered a whiff of acetophenone, the chemical that provides cherries with their sweet smell, while simultaneously receiving an electric shock. Under neutral conditions the mice would sniff and scurry around to seek out the sweet-tasting cherry, their nucleus accumbens lighting up in anticipation. But instead the mice learned, through repetition, to associate the sweet smell with the negative experience of shock, and freeze. The animals even started to sprout extra neuronal branching and new circuitry in their noses and the smell-processing parts of their brains, to help support this new behaviour. Incredibly, this learned behavioural response was then passed down to the mice's pups, *and their pups*. Succeeding generations of mice would automatically freeze as soon as they smelt cherries, even though they had never been exposed to the pairing of cherry smell with electric shock.

This finding was a revelation. How could a new memory, learned as an adult – the association of electric shock with the smell of cherries – be passed across generations? The answer is, basically, epigenetic modification. It turns out that instilling fear *did* trigger genetic changes, not in the DNA code but in how it was used in the mouse's body. The way in which the cherry-odour receptor was made, where it was produced and in what quantities were all altered. This 'flipped switch' was expressed in the mouse sperm cells, then passed on to future generations. Researchers have since built on this finding, pairing the shock with alcohol instead of cherry smell, and are subsequently seeing mouse offspring deterred rather than attracted to alcohol later in their lives. If this finding holds true for humans it could help to explain how anxieties or phobias can be transmitted between individuals even when there is no trigger, and how complex behaviours can be passed on through generations even when the offspring has had no opportunity to learn through observation.

Now, I'm not suggesting that we give ourselves mild electric shocks every time we walk past a bakery, but these results might point the way to a future in which our environment and genetic destiny could be tweaked to alter our emotional response to food, leading us towards healthier choices, and even altering our genetic responses for the benefit of future generations. As that very suggestive study using alcohol demonstrates, the applications in terms of steering us away from addictive or compulsive behaviour could be truly life-changing for millions of people.

Understanding how our preferences and appetites are pre-determined seems, paradoxically, to be opening up new routes by which we can reshape the destinies written across generations. Epigenetics is also demonstrating that evolutionary timescales for genetic change are not the only game in town any more and the interplay between the circuitry we inherit and the environment we live in is highly complex. We're only now beginning to understand it and there's a way to go before we can grasp its full potential but, given the pace of advancing knowledge, perhaps there are reasons to be cheerful about our chances of escaping the lure of doughnuts.

Epigenetics isn't the only branch of science providing hope for humanity's appetite issues. There have been other technological and knowledge breakthroughs that help to illuminate how the brain makes decisions, thinks and feels. Optogenetics, developed in the first decade of the twenty-first century, is one of the most revolutionary. So much so that I'm willing to wager that those responsible for its invention and refinement, including Ernst Bamberg, Ed Boyden, Karl Deisseroth, Peter Hegemann, Gero Miesenböck and Georg Nagel, will win a Nobel Prize. The technique uses genetic engineering to control electrical activity in the nervous system with light and allows researchers painstakingly to dissect the complex circuit board of our mind not just in terms of anatomy but also functionality. Optogenetics means that we can instantaneously and precisely switch on, or off, discrete pathways in the brain. This single technique has revealed how complex behaviours, from love to social anxiety or addiction, are directed in our brains. The way it has opened up our understanding of psychiatric conditions is unprecedented and we'll be coming back to its implications for sufferers in later chapters.

In the meantime, I want to explore a question raised by the epigenetic studies into how nutritional deficiencies before birth lead to altered food cravings in later life and future generations. Epigenetic modification might be the mechanism by which this information is transmitted, but we still don't know exactly which brain circuits support it. Could optogenetics help us to understand more?

Researcher Dr Denis Burdakov at the Francis Crick Institute in London has employed optogenetics alongside other technologies to unpick the circuitry directing our brain's response to nutritional intake. His laboratory has been investigating the region in the middle of the brain called the hypothalamus that's involved in controlling basic mammalian functions: body temperature, thirst, sleep and hunger. This is also the region where the gene FTO is expressed, a variation of which means you continue to crave food even when sated, and are highly likely to be overweight.

The hypothalamus contains specific neurons that sense the macronutrient balance in the diet. So, rather than simply monitoring the caloric content of food, they can measure dietary balance. Denis's experiments have shown that these neurons help to analyse whether a mouse's diet contains enough essential amino acids, nutrients that their bodies – and ours – require but cannot produce so they have to be obtained from food. Having completed the analysis, the cells send messages to the reward region of the brain, the nucleus accumbens, via dopamine signalling, to alter the pleasure response. This neat molecular pathway acts on the neural circuitry in a way that is similar to a double negative sentence. The end result is that if there are not enough nutrients in the available food, the mouse is primed with the motivation to seek out more. If he now has the correct amino-acid balance in his body, he will feel sated and basically stop eating. The hypothalamus will instead direct the mouse to take a nap. No further hunting for nutrients is required, so why not rest? And the FTO gene appears to be involved in this process.

Denis's findings might open up new avenues for helping us to control the way we balance the nutrients on our plates, to encourage feelings of satiety. Perhaps if we can direct the reward pathway to be driven to healthier food options, particularly for those of us with the variant of the FTO gene that inclines us to obesity, we will be in with a chance of outrunning our fate. I've personally tried to translate this research into my life and generally it seems to work for me. If I can't sleep, if I'm feeling anxious and alert but need to get to the land of Nod, I reach for late-night nibbles rich in essential amino acids, such as soy, buckwheat, quinoa, egg or chicken, and it seems to send a signal to my brain to switch off.

Looking further into the future for biologically inspired interventions, the revolutionary new findings made possible by technological advances such as Brainbow Mouse, optogenetics and genetic engineering have been combined to create a pioneering targeted technique called deep-brain stimulation, which offers a new treatment for people suffering from addiction, depression and obesity. Some patients who have not responded to other treatment regimes have elected to take part in clinical trials. Preliminary results are very exciting, but it's extremely invasive so not to be undertaken lightly. A surgeon opens the skull and embeds a minuscule remotely controlled electrical stimulation device into the nucleus accumbens, then closes the skull and stitches up the scalp. Switching on the device instantaneously activates the pleasure circuit, switching *off* symptoms. The person feels reward without the intake of whatever they're craving. An inherited predilection for sugar and fat could literally be switched on or off. There have already been some major successes in the treatment of those with heroin addiction and severe depression, and the technique is now being trialled as a treatment for Parkinson's disease, as we will see later.

Even knowing what we now know about brain functioning, it is still staggering to observe that altering the electrical current in a specific pathway of the brain can dramatically change some of the most entrenched behaviours in the human repertoire, not to mention obliterate highly distressing symptoms of disease. Deep-brain stimulation introduces the possibility that advances in neuroscientific research will start to yield more selective therapeutic applications, allowing us as a species to outrun the less welcome aspects of our biologically driven fate. There are sure to be other equally exciting discoveries and therapies coming out of neuroscientific research all over the world, but for now their applications are to be carried out by doctors once a patient has been diagnosed. Are there less invasive ways that individuals can use the new neuroscience knowledge to change their own food-related destiny before morbid conditions, such as obesity, diabetes or addiction, develop?

Outrunning our fate to be fat

One of the things I had talked to appetite-and-genetics don Giles about was whether, in his experience, people found it empowering to have their genetic predisposition to obesity tested.

What would happen if we all undertook a genetic test to find out the extent to which we're vulnerable to genetic obesity, and our triggers? Genetic testing is commercially available at a minimal cost. Would knowing exactly how our biology is paving the way to, say, morbid obesity, naturally light up our amygdala with fear, driving us to turn our backs on unhealthy food?

There have been studies looking at precisely this question, conducted by Professor Theresa Marteau and her team, also of the University of Cambridge. Her conclusion was that, in the short term, knowing that you carry the genes predisposing you to obesity can help you control your cravings and make healthier choices. Unfortunately, though, that knowledge is not a predictor of long-term behaviour change. In fact, some studies have found that once people know their biological fate, after an initial push against it, they seem to revert to their path of destiny more rapidly. It seems to demoralise them, disempowering and removing the incentive to fight it. 'It's not my fault, it's my genes' is the perfect excuse for lackadaisical behaviour.

If that leaves you feeling demoralised, please don't despair. There's powerful biology at work every time we order in a café or wander through a supermarket. Merely knowing that is rarely enough to bring about change but it can be a helpful spur to take up a range of strategies for enacting your own behavioural tweaks. After all, plenty of people acquire the taste for different foods, alter their diets and lose weight. With better knowledge of how we personally respond to motivation and reward, it becomes possible to make tweaks in our environment to nudge us towards making more healthy 'choices'.

We can, of course, carry this out by gaming our reward systems with alternative treats, teaming up with buddies for mutual encouragement and accountability and implementing super-practical policies, like never having our trigger foods in the house. The fact is, though, that meaningful improvement for society as a whole will come by tweaking the public environment — we will return to this when we look at how the insights of neuroscience can be applied to bring about changes in diet and food choice at population level. This might feel like the nanny state interfering in very personal issues, but as we start to appreciate the limits of our individual autonomy, we might begin to favour such policies.

I have a habit of asking the research scientists I meet how they apply the often highly theoretical fruits of their labours in their own lives. I asked Giles — he of the slightly gloomy conviction that humanity was born to eat doughnuts — what his own tactics were in the fight against obesity.

He admitted that, despite the research indicating that knowledge of your genetic disposition is ultimately unhelpful, he has examined his own DNA. It turns out he is among the 25 per cent of the world's population who carry the FTO gene variant, veering him towards obesity. He works hard to ensure he eats nutritionally rich foods, laden with essential amino acids so that his hypothalamus provides a firm nudge towards feelings of satiety. He also *always* takes the stairs rather than the lift. And — this is the crucial bit, he maintains — he is aware that, with his Chinese heritage and familial prior exposure patterns, he is drawn more towards salty and fatty foods, which he cannot do much about. His wife, on the other hand, has a weakness for chocolate. Between them they've come up with a household policy: nothing resembling pork scratchings, salami, crisps or chocolate is allowed in the house. They support each other's mutual goal to maintain a healthy weight, in the hope that they will pass on healthy eating habits to their children. Giles and his wife are making conscious efforts not to tread their predestined path. They are drawing on the power of a region in the brain that plays a role in social interactions, applying mutual peer pressure and tapping into their playful and competitive characteristics to keep themselves trim. And it's working for them.

So where does this leave us as we struggle to come to terms with the suggestion that we're nowhere near as free as we assume to make even seemingly basic decisions about our lives?

By the time I'd finished talking to the appetite scientists I had been forcefully reminded that, when it comes to human brains, *nothing* is basic. Even a supposedly low-level behaviour like eating is a highly tangled web of inherited preferences, preferences learned in early years, epigenetic feedback loops and plain old instinct to seek out high-calorie foods and keep eating them. But the central paradox is that even with a behaviour as driven by universal circuitry as food preferences, it is also composed of so many highly intricate and complex parts that every one of us has a unique set of behaviours around food. So, no single strategy for behaviour change will help everyone to lose weight or eat more healthily. We are all creatures of appetite driven by our own particular cravings, but if you take the time to find out what works for you, change is possible. After all, even Giles, who knows more than most about how genetics influences people's weight destinies, tweaks his own food choices successfully, knowing that his faith in the possibility that his behaviour changes will last is at least as important as any individual strategy he could employ.

The neuroscience of fate is a curious and paradoxical field of study, then, which is perhaps only to be expected, given the mysterious and awe-inspiring complexity of the organ at its centre.

CHAPTER FOUR

The Caring Brain

You probably won't find it particularly challenging to your assumptions to be told that when it comes to sex human beings are hard-wired to behave in certain ways. As a species we reproduce sexually, and although we have a lot of non-reproductive sex and not all human sexualities are geared explicitly towards reproduction, this fundamental biological truth drives and shapes behaviour.

Our culture has long viewed sex as an unruly force. Even once we'd discarded the idea that it was inherently sinful we persisted in thinking of it as the product of subconscious desires and repressed emotions. In some sense we're used to thinking of ourselves as being not fully 'in control' when we're engaged in sexual activity. That's partly the appeal. It's an opportunity to relinquish at least some of the conscious, analytical agency of our daily lives. But what if it isn't just sexual choices that are driven by powerful instinct and the lure of the brain's deliciously rewarding pleasure chemicals but also romantic love, partner bonding, parenting, friendship and wider social bonds?

As new technologies allow us to peer into the brain we are starting to see how all forms of human connection are driven and controlled by deep-brain functioning. Love, it seems, is largely a by-product of the brain circuitry that prioritises reproduction and the survival of our species. It's fundamentally linked to the functioning of the reward system, which we encountered in the previous chapter and will see cropping up throughout the book: its role in driving behaviour of all kinds is so crucial. This chapter sets out to show that the singer Robert Palmer got it right: people really are 'addicted to love' – of all kinds. Our brains crave romance, affection and social bonding, all of which drive us towards forming relationships.

The science of the drive to reproduce probably feels relatively familiar to many of us. Quite apart from the dimly remembered biology lessons of our schooldays, research into sexual behaviour is a boon to writers of newspaper features. Sex sells, and so does the neuroscience of sex. For that reason, any study of sexual behaviour or difference between the sexes that draws on neuroscience or neuropsychology tends to generate more than its fair share of neurohype and pseudoscience. Setting aside the more outlandish claims, there is no doubt that sexual behaviour is shaped by drives and cues of which we may not be consciously aware. There have been a number of studies, for instance, which conclude that heterosexual men will rate a woman's voice and dance moves as more attractive during her five-day highly fertile window. This effect can even be monetised. A study analysing 5,300 lap dances performed during different stages of the menstrual cycle revealed that dancers will get almost double tips on the days they are highly fertile compared to when they are menstruating and pregnancy is very unlikely. Their customers are unaware of this, of course. They think they're objectively evaluating how attractive the dancer is.

When it comes to sex, it seems that a choice that may feel highly personal and deeply intimate – something as basic as whether to have sex or not – is to a large extent the behavioural result of our brains' coding to seek maximum opportunities for our genes to be passed on.

The ramifications of such research extend beyond the realm of sex to reveal a huge amount about how intimacy, trust, affection and love are also, to an extent, the products of deep-brain functioning. We are predisposed to crave intimacy and affection, and that need, whether it's expressed as romantic love, parental devotion or the affection we feel for our close friends or even our social group, is generated by the same mechanisms that send us out to reproduce and have fun trying. Romantic love is in some sense the conscious brain's apprehension of these mechanisms, and social and emotional bonds of all kinds are deeply rewarding to most people on a neurobiological level. Positive social interaction triggers dopamine release. (As ever, and you will grow tired of this caveat if I'm not careful, the extent to which this is true of any individual varies greatly.)

How sex drives us to love

Let's begin with sex, since it's at the root of so much of this behaviour. More specifically, let's consider pleasure. The function of orgasm for both men and women is partly to encourage us to engage in sexual activity. Thanks to studies on this subject conducted since the seventies, we know that male and female orgasms have similar electrical effects on the brain. More recent imaging technology has enabled researchers to analyse the brains of volunteers as they bring themselves to climax while lying in a brain scanner. The scans revealed that orgasms ignite the neural circuits of pleasure. An orgasm is produced when a build-up of dopamine is released suddenly to activate the nucleus accumbens and produce intense feelings of pleasure. It's the reward circuit again. In a crucial sense, an orgasm takes place not in the genitals but in the brain. In fact, even the anticipation of sex can trigger dopamine release to the reward circuits, which can be enough to get us seeking actual sexual activity.

So far, so good. Sexual intercourse is still the easiest and most common route to reproduction and at least in theory, though not always in practice, it is highly pleasurable. We have sex because of our drive to pass on our genes and our fondness for pleasure. Except that sex is more complicated than that and our sexual behaviours are influenced by dozens of factors, from our self-identity, our sexuality, our gender alignment and our repertoire of experience to our age, class, state of health and so on.

My focus is to use insights from neuroscience to unhook sex from reproduction and, indeed, from pleasure. I take it for granted that those two drivers, those pre-wired imperatives, compel us in ways of which we may or may not be fully conscious. What, though, can neuroscience tell us about people who do not want to have sex, who define themselves as asexual? Or about homosexuality? Is there any evidence to suggest that different sexual orientations are determined by our inherited genetic legacy or by deep-brain functioning rather than our personal preferences built up from experience?

Before we move on to look at the neural basis of non-reproductive sexuality and asexuality respectively, let's examine how little conscious agency heterosexual people frequently have in their choice of mate. The fairy tale of the quest to find our One True Love has taken a bashing of late, and not before time, but the idea of The One still exerts a powerful hold over our collective imagination. Many of us contort ourselves in working out which personal qualities are non-negotiable in our life partner and apply pop psychology to ourselves and others in pursuit of the perfectly honed dating strategy. We rationalise our love life over a couple of beers, and come up with all kinds of stories to explain our successes and failures. In other words, sex and love often feel like some of the most analysed aspects of our lives. Whereas in reality there is plenty of evidence that our choices are driven by biological imperatives of which we are not consciously aware, such as 'genetic hunger' for complementary genetic material to pass on to any possible offspring.

An interesting experiment, first performed by Claus Wedekind of the Zoological Institute at Bern University and subsequently replicated in the United States, showed that women subconsciously sniff out their preferred partner as part of their mating-evaluation repertoire. Researchers asked a group of men to wear a T-shirt for a couple of days without washing, using deodorant or eating or drinking anything too distractingly smelly. They then asked a group of women to sniff the T-shirts without knowing anything about the wearers and rate the scents for attractiveness.

It transpired that the women much preferred the body odour of men whose immune systems were very different from their own. The difference lay in the hundred or so genes known as the MHC, the major histocompatibility complex, which is code for proteins that help the immune system recognise foreign bodies, including pathogens. These genes play a dual role in determining how you smell and your immune-system composition. A mate with different gene variations from your own would produce offspring who are resistant to a broader range of infections, boosting their chance of survival. The women could literally sniff out Mr Right, with optimum genetics in mind. It appears to be a completely unconscious behaviour, written into their genes and wired into their brain. Incidentally, this olfactory ability is less pronounced in men, which suggests that men are generally less sensitive to smell than women and have less invested in sniffing out the 'right' partner: they do not make the same sacrifices of time and energy in bearing and bringing up children as women do.

Interestingly, the researchers also discovered that the contraceptive Pill, which hormonally emulates a perpetual state of pregnancy and thereby renders the woman temporarily infertile, turned these results on their head. Women on the Pill were more likely to approve of the smell of men who were genetically similar to them in terms of immune-system composition. In essence, if you are on the Pill you probably prefer the smell of men who are genetic relatives, such as brothers or cousins. After all, if you *were* pregnant, having some male relatives around to protect you and your child would be very handy. Other studies have indicated that hormonal contraception can alter the wiring of the brain to affect choice of boyfriend, which poses the question of whether a woman who comes off the Pill to get pregnant might discover she's no longer attracted to her partner.

But before any reader starts to panic about whether her taste in men might change drastically in such circumstances, it's worth sounding a note of caution. Given that any individual's MHC profile (a crucial piece of genetic coding for the immune system) is as singular as their fingerprint, it's highly unlikely that coming off the Pill will reveal that you're about to try for a baby with somebody whose profile is too similar to your own. To put this in reassuring context, Carol Ober, a geneticist from the University of Chicago, carried out a study within the 40,000-strong Hutterite community in the rural Midwest, looking at genetic compatibility in couples. The Hutterites are an ethno-religious group not dissimilar in beliefs and practices to the Amish. They are forbidden to marry outside the clan, but even within this much reduced gene pool, most couples' genetic compatibility seemed to pose no significant threat to their offspring.

Some studies have suggested that the more similar a couple's genetic profiles the harder it is to maintain long-term attraction, so this might be more of a worry for people either using the Pill or in a relationship with someone who uses it. Perhaps you're actually too similar and are fated for passion to fade? There's no doubt that sexual attraction serves in part to boost immune-system diversity in any future offspring, and that this neatly illustrates how important life decisions are shaped by forces outside our conscious control. That said, it's probably safe to ignore any newspaper headlines about the 'divorce pill' for now. Long-term attraction isn't easy in any circumstances, and there are almost as many factors that contribute to a relationship standing the test of time as there are fish in the sea, or MHC profiles. Interestingly, there is a recent study that could be used by the commitment-phobe among us to explain their behaviour. Researchers from Texas University analysed brain tissue taken from ten different species – five typically monogamous ones, along with five of their more promiscuous relatives – and identified twenty-four genes whose activity was consistently dampened down or heightened in the different groups. Admittedly, while the species investigated did not include humans, the research does point towards a genetic trick that has been evolutionarily conserved within these species that might help child rearing couples stay together.

The first flushes of romantic love appear to be a side effect of the basic reproductive urge. Plenty of studies have demonstrated that the euphoria of falling in love is the result of a cascade of brain activity that results in your concentrating all your reproductive attention on one specific promising prospect. Similarly, some studies have shown that a typical life cycle for a romantic heterosexual relationship is approximately seven years, or long enough for a partnership to form, bed in, produce children and nurture them through the earliest, most vulnerable years.

As always, though, it's a mistake to draw simplistic conclusions. This might well be a species-level truth but it can't be applied at an individual level to explain the choices of any single person or couple. Reproduction is a strong driver, as is the human preference for novelty. Does this mean we're hard-wired to commit adultery? You can argue that human beings are inclined to this behaviour (and plenty of people have). But, as it turns out, a reliable source of intimacy is also highly motivating. It's just as possible to suggest that we're built for long-term love. Your brain has evolved to encourage sustained intimacy with a single individual: regular rewarding intimate interaction can be enough to keep you hooked for life. Studies have shown that long-term couples who describe their relationships as happy seem to be physically addicted to each other. The mere thought of their partner can activate their pleasure system, lighting up the brain with dopamine in a manner resembling an addict's anticipation of their drug of choice.

A number of neurochemicals are involved in maintaining our relationships beyond the first throes of passion. For example, the tender touch of a partner can stimulate nerve endings in the skin to send electric signals to the hypothalamic region of the brain, where the pro-hormone oxytocin is released. Oxytocin is heavily involved in bonding between individuals and is particularly crucial in bonding a mother and her newborn. It's powerful stuff, acting in a similar way to alcohol to activate inhibitory nerve cells in the prefrontal cortex (an area involved in decision-making) and the limbic system (governing your motivation, emotion, learning and memory). By activating these inhibitory nerve cells it dampens stress and anxiety and puts the brakes on social inhibition. This helps to enhance feelings of wellbeing, relaxation and trust, thereby increasing your chances of reaching sexual climax. In fact, oxytocin has been used successfully in couples therapy, as a nasal spray, to bring about feelings of closeness.

Oxytocin is not all sweetness and light, though. It has been somewhat cloyingly termed 'the cuddle hormone', but as well as enhancing pair bonding it also promotes feelings of territorialism and aggression towards outsiders. This has a benign manifestation in the common post-coital urge to hide away with a lover and ignore everybody else in the world. That romantic feeling of caring nothing for anyone or anything except the beloved is in part the result of oxytocin. Less charmingly, it's been shown that oxytocin is also implicated in the dynamics of social hostility. It encourages individuals to favour their in-group strongly, at the expense of others.

If romantic love and long-term pair bonding are to a certain extent generated by the release of pleasure chemicals in our brains, to what extent do we consciously choose to stay in a successful relationship? Does it take something from the awe we feel for those amazing couples who've been together for fifty years, and still seem to delight in each other's presence, to know that on one level it's down to their brains' dependence on the dopamine hit they provide for each other?

I prefer to find it awe-inspiring that human brains are so cunningly designed that they're capable of such things. For me, this knowledge, far from stripping away the mystery of long-term intimacy, can be empowering and useful. If we know that neurochemicals have a crucial part to play in the durability of long-term relationships, we're in a position to look after our own relationship's health. We can maximise our efforts to trigger their release by regularly engaging in mutually pleasurable activities with our partner. That may be sex but could just as effectively be massage, hugs, caresses as we pass them in the house and simple acts of kindness. Studies have shown that the mere act of asking your partner about their day, then listening to what they say and commenting with sympathy is, perhaps unsurprisingly, enough to trigger and strengthen the bonding process.

So much for the ramifications of the drives to pass on our genes and seek out pleasure for heterosexual love. What about sexual behaviours that can't be so easily tied back to an initial 'genetic hunger' as a determining factor in partner choice? What, if anything, can neuroscience tell us about homosexual orientation?

Even to ask the question means we must acknowledge the risk that the study of behaviour, especially sexual behaviour, can be appropriated for ideological ends. Some extreme social conservatives have attempted to use science to justify their argument that any form of sexuality outside the heterosexual 'norm' is 'deviant', an argument that is as flawed as it is objectionable. As a biologist you quickly get used to observing the staggering variety of behaviour exhibited by any single species. The great truism that all behaviour sits on a spectrum is nowhere more true than when it comes to sexuality, and though I wish I didn't have to make the point, because to my mind it's blindingly obvious, there is no such thing as normal when it comes to human sexuality or, indeed, to the human brain.

With that clarified, let's turn to a highly reliable source of information about the neurobiology of sexuality. The pristine new building of the Medical Research Centre's Laboratory of Molecular Biology is known in some quarters as the 'Nobel Prize Factory'. I went to pay a call on Dr Greg Jefferis, who is researching the differences in sexual behaviour of male and female fruit flies. And before we go any further, I know it might seem bonkers to suggest that we can learn anything about human behaviour from fruit-fly sexuality, but studies on model organisms such as fruit flies and mice have been yielding breakthroughs for years in this area, as in the study of appetite. You can't make simple extrapolations but there are enough similarities in certain brain structures and functionality to ask and test interesting questions.

Greg has painstakingly mapped the brain circuits that are active when a fruit fly is engaged in courting behaviour. He has isolated a three-nerve-cell sequence that detects male pheromones and seems to make females more willing to mate while having the opposite effect on males, who become more aggressive. By puffing pheromones at the flies, then watching to see which cells lit up, he could determine that it's the last cell of the three that acts as a switch in the brain. The switch is 'controlled' by a single gene that is normally only expressed in the male. Other fly researchers have used genetic engineering to express this gene in female flies. After this manipulation, instead of responding to male advances the females tried to mate with other female flies. The flies' sexuality had been altered at the flick of a single-gene switch. Scientists at the lab of Catherine Dulac, Higgins professor in molecular and cellular biology at Harvard University, have been able to produce similar 'sexuality switches' by manipulating the pheromone-detection system of mice.

'There are clear parallels between fly and mouse pheromone work and I would expect to see similar kinds of brain-circuit dimorphism in future studies of the mammalian brain,' Greg told me. 'But human sexuality is infinitely more complicated than this, with many layers of neural control that must be influenced by social and developmental factors. Nevertheless, I would be very surprised if there were not a genetic contribution to the fact that the majority of humans choose an opposite-sex partner.'

As with all complex behaviours, it won't be the case that a single region is 'in control' of sexuality. Neither are we close to identifying the genes that are implicated in sexual orientation. Decades of research have attempted to ascertain which, if any, genes are associated with either male or female homosexuality, and if you trawl through the studies you'll see pretty inconsistent and generally inconclusive results.

That said, there is some evidence to suggest that, as with fruit flies and mice, sexual orientation in humans may have a neuro-biological component. In both men and women, the number of older same-sex siblings have been found to correlate positively with homosexuality: each additional older brother increases the odds of a man being homosexual by 33 per cent. But is this due to any innate trait or is this behaviour somehow learned through the life experience of having older same-sex siblings?

Some researchers have suggested that it is inscribed into the developing foetal brain through the reaction of the mother's immune system. For the men, for whom there is more data, the hypothesis goes that each male foetus provokes a maternal immune response directed to attack the male sex hormones produced in her body, which are regarded as 'foreign agents'. With each successive son the mother's body is able to mount a faster immune response, maximising the window of opportunity to 'feminise' the embryo's brain by reducing its production of testosterone. It's this lower level of testosterone production that has been shown to correlate with a person identifying themselves as homosexual.

This research is speculative and certainly doesn't serve on its own to account for every instance of homosexuality. For a start, there are plenty of homosexual men who are the oldest child in their family, and this mechanism is unlikely to be a contrib-uting factor to female homosexuality because a female foetus does not provoke the mother's immune response to such a strong degree.

Greg's research into the neural basis for sexuality in fruit flies is a painstaking demonstration that, in this species at least, even the seemingly immutable alignment between biological sex and mating behaviour all comes down to a circuit in the brain, but it would be stretching a point to extrapolate from there and suggest that we are now close to understanding the neural basis for human homosexuality.

Research into the asexual brain, for now, is at a very early stage. Defining what we mean by asexuality yields different insights. Associate Professor Lori Brotto at the University of British Columbia has conducted studies with volunteers who describe themselves as asexual but retain their libido and simply don't want to act on it with anyone else. She looked for biological markers of this behaviour and found that generally such people were around 2.5 times more likely to be left-handed than their heterosexual counterparts. They were also more likely to have spent either more or less time in utero than the typical forty weeks. Lori believes these findings point towards an early neural wiring difference that might contribute to their later asexuality.

A lack of interest in sexual activity is frequently a temporary state of affairs triggered by specific circumstances in an individual's life, rather than a long-standing trait. Reasons abound, from illness or trauma to the more benign, such as just having had a baby or being stressed from work. Nevertheless these contextual factors have a neural component that will impact on a person's desire to engage in sex. Libido is, after all, in some sense a neurochemical phenomenon. It is pretty unanimously agreed that for any sexual act to be initiated complex interactions are required between multiple neurochemicals acting on brain and body. For example, dopamine, oestrogen, progesterone and testosterone play an excitatory role, while serotonin and prolactin are inhibitory. The levels of these neurochemicals in an individual might be dictated by genetics to some degree, but they are also sensitive to any number of environmental triggers.

So, for a long time after the birth of my son I was not interested in having sex with anyone. Even if Fate had set me on a star-crossed path to bump into Ryan Gosling I would probably have sat him down for a nice cuppa and a chat. While I was breastfeeding my son, the levels of prolactin hormone in my body were high, inhibiting my sex drive. I suspect that if you had put me in a brain scanner during the first chronically sleep-deprived year after his birth and asked me to ponder sex, you would have seen my amygdala region light up with a fear response rather than the more pleasurable association of dopamine rushing to the nucleus accumbens. My resources were consciously and unconsciously focused on looking after the child I already had.

Although results from the research into the neural basis of non-heterosexual behaviours are still emerging, sexual relationships, in all their forms, reproductive and recreational, are driven by brain circuitry and instincts honed over hundreds of thousands of years. Sex is one of the behaviours that most intimately connect us to non-rational drives, but it is never 'merely' about subconscious motives. It is also something we engage in with the fully conscious part of ourselves, in the context of our own life story and the social norms that establish codes for 'acceptable' sexual behaviour. In the UK, homosexual activity between men was only decriminalised in 1967 but is now widely accepted. New adoption rights and breakthroughs in IVF provide more possibilities for non-standard families to emerge. Increasing numbers of transgender men are giving birth. Simultaneously there is heightened awareness that our ever-expanding population may be exceeding the planet's capacity to support us all, while financial pressures are forcing many people to re-evaluate the 'need' to reproduce.

So, sex in whatever form, and even the lack of it, is in some sense socially driven. Accepted norms and external pressures may compete with desires that are deeply biologically ingrained. While it is currently impossible to tease apart the impact of these factors precisely, what can be said for sure is that some of the most precious aspects of our unique personal lives derive from our universal sexual imperative. The physical longing we experience when we fall in love, the fierce urge to nurture and protect our loved ones, even the jealous hostility towards anyone who threatens our love bond, all these vital emotional states are the result of intense neurochemical activity that has evolved over millennia to encourage us to invest our time and energy in sex and parenting.

The surprising truth about the nurturing instinct

Parenting among human beings is, like sex, overlaid with layers of socially constructed behaviour but most of us are probably comfortable with the idea that its primary motivation is unequivocally biological and innate. Once an animal, any animal, has successfully reproduced, the offspring must be nurtured to sexual maturity so that they in turn can pass on the precious genes. The love a parent feels for their child is, on some level, driven by wishing to see them survive and reproduce. In fact the parent–child paradigm is a particularly useful prism through which to view the notion of love as a behaviour to which we are predisposed and unconsciously driven. It allows us to see very clearly that affection is in some sense the by-product of an innate drive connected to both sex and reproduction. By and large (there are exceptions) we are fated to nurture and love our children, though they may or may not love us back.

Some fascinating research has investigated the mechanisms by which this nurturing behaviour is switched on in new parents. Male mice have a reputation for being not so much 'hands off' parents as 'teeth on'. They will generally, and rather gruesomely, kill and eat newborn male pups when they stumble across them, presumably to get rid of any future competition. However, three weeks after the male mouse has engaged in sexual activity there is a window of time during which he will exhibit nurturing behaviour towards any newborns he finds. Catherine Dulac of Harvard University demonstrated that for a short period he will groom them, pick them up in his mouth and return them to a nest. The gestation period for mice is – you guessed it – three weeks. It seems there is an internal clock within the male mouse brain that flicks off typical aggression and replaces it with caregiving at precisely the right moment to nurture any of his potential offspring. The mouse is uncertain whether or not the pups are his. It's not as if he's been hanging around in some mouse equivalent of the traditional nuclear family. He just indiscriminately looks after any pups he encounters, even male pups, if their time of birth correlates with his previous mating session. The exact mechanism governing such an abrupt alteration in behaviour is still not clear. It may be that hormones released during sex, including a reduction in testosterone, induce the formation of new brain circuits to instruct the paternal behaviour.

A similar, albeit thankfully less cannibalistic, trend has been observed in human males. Testosterone in expectant fathers will drop by a third, while the hormone prolactin, connected with nurturing behaviours and lactation in pregnant women, will rise significantly in the weeks before the baby's due date.

It's interesting to conceive of a chronological control switch in our brains, instructing even complex behaviour, such as nurture. As we move from the wider point about parental love to the specifics of a mother's role versus a father's, it feels wonderfully refreshing. Some people seem to believe that a mother is hard-wired to devote herself to her offspring, while the father is innately wired to be more hands off.

Such stereotypes can be unhelpful on a number of levels and it's obviously not just academics of every discipline who love to look for evidence to help reinforce or challenge these misconceptions. In pubs and family kitchens around the world there's no shortage of opinion on what's 'only natural' for men and women alike. Even in the avowedly egalitarian societies of the post-industrial world, where the majority of women now work outside the home, there is still a great deal of cultural anxiety about the role of mothers and fathers as parents.

As with sex, we should definitely be wary of essentialist arguments about parenting. Neuroscience has been co-opted by some writers to bolster socially conservative arguments about women's roles, and mothers' roles in particular. For a rigorous and often funny demonstration of why most of it is pseudoscience writ large, Cordelia Fine's *Delusions of Gender* is unbeatable and not to be missed by anyone who is sceptical about neurohype in any sphere.

In one study, cited by Fine, necessity is shown to be the father of invention. Male rats don't typically get proactively involved with infant care, but if you leave a male rat in a cage with a newborn and there's no mother to take care of it, the male rat proves perfectly competent at grooming, nurturing, even nest-building. It takes a couple of days but before long the male is snuggled up to the baby as if he was born to caring. As Fines puts it, 'The parenting circuits are there in the male brain, even in a species in which paternal care doesn't normally exist. If a male *rat*, without even the aid of a William Sears baby-care manual, can be inspired to parent then I would suggest the prospects for human fathers are pretty good.'

It seems that parenting, in either biological sex, results from a deep-seated innate drive that helps to bolster affection and nurturing. Genes, hormones and environment are all crucial, and unless you assess them all together, you can't say anything reliable about behaviour. Male rats are not paternal, except when they have to be because of environmental constraints, in which case environment triggers hormones that activate neural pathways that lead to behaviour change. A similar switch occurs in male mice to bring about paternal-care behaviour, the trigger in this case being time. It's too easy to let received ways of thinking blind us but it's more interesting when we don't.

Affection is a neurochemical event, in humans as it is in rats. It's motivated by reproduction and also by the reward circuit: nurturing is not only necessary for survival but deeply pleasurable.

Pleasure, personality and the social brain

Moving from one-to-one romantic relationships and the bonds forged in parenting to the dynamics governing friendship and kinship, what can neuroscience tell us about these forms of affection that we might think of as primarily social constructs? The fact that studies have linked oxytocin production to feelings of both romantic infatuation and hostility to outsiders is one surprising insight into the links between our neural and hormonal activity and our social bonds. I wanted to know more about how we evolved the ability to form close cliques of friends alongside looser social connections within our community. Most people, even the relatively shy or introverted, have a staggering capacity to navigate our complex social world, and much of this functioning takes place with seemingly unconscious ease. This suggests that all relationships are to some extent reliant on deep-brain functioning to do with pleasure, reward and motivation.

I spoke with Robin Dunbar, professor of evolutionary psychology at Oxford University, who has spent his career researching the social brain. Robin has explored how – and, crucially, why – human beings seem to be biologically driven to find the process of reciprocating affection and attention rewarding and how this, ultimately, helped us to evolve as a species. He talked to me in his living room, spring sunshine streaming through the bay windows, his enthusiasm for his subject shining through in his voice and the brightness of his expression.

Robin is clear about the benefit of friendship to our personal wellbeing. 'It's the single most important factor influencing health and happiness,' he insisted. Affection for, and engagement with, another person are basically good for us. He told me about his favourite study on the subject, which showed that the best predictor of whether or not you will recover after a heart attack is not whether you give up your twenty-ciggies-a-day habit or renounce your cholesterol-dripping fry-up each morning, but the strength of your support network and friendships. Physically affectionate touch, such as hugging, expressions of concern and laughter all boost the production of endorphins, which have a positive effect on our immune system and lead to faster recovery times and greater resistance to infection as well as better mood.

'Friendship is a self-protection mechanism built into our brains,' he explained. 'Your friendships help buffer and protect you during times of need, but alliances have to be set up well ahead of when you need them, and this takes energy. Making friends and main-taining them is cognitively costly. But friendships should be worth this investment, and at times you have to place the value of friendship and community over immediate self-gratification.'

Robin believes that we evolved this ability to plan for our futures by building and nurturing supportive friendships at the same time as humanity developed its orbital prefrontal cortex (the bit of your brain immediately behind your eyes, which is involved in inhibiting impulses and processing emotions). He believes that its relatively large size in humans is the key underpinning to our evolution as a species. It's not our phenomenal processing power *per se* that is important, more our ability to successfully maintain and navigate our way through large and complex networks of relationships. Robin has analysed what he calls 'the social-brain hypothesis' on a species-by-species level, and corroborated that the relative size of this brain region in mammals, and particularly in primates, correlates with the size of their social groups. It peaks with us humans, where he suggests that we are able to maintain stable social relationships with, on average, around 150 people. Beyond this approximate size of community, which is determined by a cognitive limit, more formal rules are required to govern social dynamics. Robin has described 'Dunbar's number', as it's now known, as being the number of people you might conceivably make an effort to contact at least once a year.

There is obviously scope for variation around this number. As with any behaviour, there's a sliding scale of sociability among us and this is where the research gets fascinating: the size of an individual's orbital prefrontal cortex can be used to predict, with a high degree of accuracy and sensitivity, the size of their social network. It appears that there are two distinct types of social temperament out there, and two distinct brain profiles associated with them. Some people, let's call them 'extroverts', have a larger volume of orbital prefrontal cortex, and they participate in a correspondingly greater range of social networks. They seem to sacrifice a little bit of quality by devoting less time to each individual relationship, spreading themselves more thinly and using their spare capacity to invest in maintaining a larger network. Other people, let's call them 'introverts', will have a smaller social network but each friendship within it will be stronger, more reliable. And, generally speaking, whichever end of the spectrum a person tends towards, they will still find it rewarding on a brain-chemical level to spend time nurturing their friendships and alliances.

These different socialising styles support a social structure that benefits the species as a whole. The smaller groups of closer friendships formed by introverts create secure pockets of social cohesion and the extroverts create bridges between the cliques, allowing ideas to be exchanged, preventing echo chambers and facilitating information and idea transfer across disparate groups. It's this capacity for variation in social behaviour – our potential as a species to forge close alliances and support each other during times of need yet also process new information and take on board different perspectives – that appears to have been a key factor in our evolutionary success. It has been crucial in producing an effective collective consciousness, as we will discover in later chapters when we uncover the neural basis for the creation of our belief systems.

Investigation into friendship and the formation of social tribes indicates that there are six key neurochemicals that help drive our social behaviour: Beta-endorphin, serotonin, testosterone, vasopressin, oxytocin and our old friend dopamine. They interact in complex ways in the social-brain network (and many of them also cropped up in relation to sexual activity) but, very broadly speaking, dopamine seems to supply a buzz of motivation and reward, while endorphins give you the comfortable, contented, relaxed side of sociability.

Robin has been particularly interested in the role of B-endorphin ever since he discovered that your brain lights up with B-endorphin activity when you laugh with others, hug, sing, tell stories, dance or do synchronous exercise with them, and that these activities all boost feelings of social inclusiveness and cohesion. 'Singing is the best for this,' Robin told me. 'A single one-hour singing session will have complete strangers telling each other their life stories afterwards.' Perhaps this also helps to explain why mass exercise activities, such as marathons and triathlons, are becoming so popular across many of our densely populated yet anonymous cities. Given that we are suffering a crisis of loneliness rivalled only by the obesity epidemic for its potential impact on health and wellbeing, the implications of this research could be profound. With a bit of imagination, maybe story-telling or a community singalong could be staged as a public-health measure, or even a reconciliation technique.

Interestingly, extremely sociable people have many more B-endorphin receptors in their orbital prefrontal cortex than those who are less socially active. Robin speculates that if you are born with a large number of these slots in your brain, to absorb endorphins, it takes more social stimulation – lots of friends – to fill them so that you get your pleasure kick. If you have only a few receptors, you require less social interaction to reach the same level of satisfaction.

So, if the number of individuals within our social circle and the style of bonds we form are largely predetermined, what about the criteria we use when choosing our friends? Is that also ingrained by factors outside our control? Robin suggests that we each have the friendship equivalent of a supermarket barcode that advertises certain key characteristics. It is activated primarily by our use of language, which makes it 'scannable' by others when we interact with them. The more characteristics we share with another individual the higher the chance of us hitting the loyalty-card jackpot and forming a close friendship. Factors include where and how we were brought up, our educational experience, our hobbies and interests, our worldviews and sense of humour, all of which are scanned via conversation. A large part of the preliminary information is imparted through our 'dialect': how we express ourselves. This allows the person with whom we're interacting to evaluate whether it's worth investing time to solidify the relationship. This process is, of course, a two-way street, so we're all constantly trying to figure out whether someone's perspective on the world will be so wildly different from our own that reaching a mutual understanding would take too much mental energy. If so, the proto-friendship will, in all likelihood, just slip away. It's an efficiency mechanism, basically, short-circuiting the need to invest huge quantities of time and cognitive capacity in finding out whether or not each individual you meet is a potential ally. And all this is happening at a subconscious level, don't forget, which helps to explain that none of us is without bias when we form evaluations of others.

Our inclination to form friendships seems to be based largely on commonalities, though hopefully not to the point of being solely drawn to identikits of ourselves. We are, after all, also built to find novelty thrilling, as we will see later. This broad preference for sameness among friends contrasts with the criteria we employ when looking for reproductive partners. The strong innate drive to produce offspring with a robust immune system has been vindicated by recent research into the crucial part an individual's immune system plays in their health outcomes. The 'inflamed-brain' hypothesis, for example, suggests that depression may be caused partly by inflammation in the brain caused by a faulty immune system. The benefit for an individual in terms of the survival of their genes is clear.

The link between social rather than sexual behaviour and immune-system functioning is illustrated by the virtually universal experience of feeling blue when suffering from a severe cold, flu or stomach illness. When our immune system is busy fighting off an infectious agent, we often feel in a low mood and typically want to limit our social interaction. Taking a duvet day or two helps to prevent the spread of the agent across our contact group. Systems across our body and brain work collaboratively to an extraordinary (and unconscious) degree for the benefit of our species and individuals alike. Our genes, behaviour, reproductive fitness and sociability seem, once again, to be entwined.

Sociability seems to be the outcome of complex mechanisms operating in tandem. It is essentially a form of insurance that depends on an ability to plan for the future, which fortunately (given how much time and energy it takes) does not require your full conscious effort. Building successful relationships based on trust, reciprocity and obligation that also allow the input of fresh ideas and perspectives is hard work, but also beneficial for individuals, their tribes and the species as a whole. So much so that it seems we are hard-wired to treasure our close social interactions and fear social rejection. Professor Naomi Eisenberger at the University of California, Los Angeles, and her colleague Dr Mary Frances O'Connor, have discovered that the impact of social rejection on the brain resembles that of a serious physical blow. 'The social attachment system is piggy-backed onto the physical pain system to make sure we stay connected to close others.' It's tempting to invoke this mechanism to explain the old adage of dying from a broken heart. The grief at a break-up with a lover, partner or a close friend can trigger the release of stress hormones and heightened activation of the pain pathways at levels and durations beyond that which an individual can endure. Chronic isolation and loneliness are strongly associated with increased rates of illness and death. That broken heart might not be merely a metaphor.

But within the range of sociability there is a category of people who are uninterested in social interaction, who are in effect 'asocial'. I'm not talking here about people who are actively hostile to social demands, or those who have a diagnosed condition, such as autism, that sometimes presents as a lack of interest in socialising, or a struggle to be sociable. I'm thinking of people who actively choose to spend most of their time alone, who are indifferent to the pleasures of social contact. How to account for such outlier behaviour when the growing body of research suggests that human beings are predisposed to seek out social contact in pursuit of the pleasures of affection and connectedness?

To investigate 'asocials' and their relation to the idea that communication, reward and sociability are innately wired (and deeply enmeshed) in our species, I wanted to explore – bear with me here – the behaviour of a bee colony. I spoke with Dr Gene Robinson, director of the Institute for Genomic Biology, Illinois University. His laboratory looks at how gene activity and social behaviour are linked in the Western bee, *Apis mellifera*. Social genomics is an emerging field of research that examines why and how different social factors, such as stress, conflict, isolation and attachment, affect the activity of the genome within the brain of an individual, and how that can shape emotions and drive social behaviour. Essentially, the objective is to understand how the environment activates genes. But could any of Gene's work have relevance to humans? Are bees and humans really comparable?

Gene, who fires out highly technical answers to my questions at an exceptionally fast, unblinking rate, agreed that it would be problematic to make simple analogies. 'But there is a region in the bees' brains called the mushroom bodies' – a clear contender for the oddest name in biology prize – 'that is responsive to social stimuli and is receptive to a neurochemical called octopamine.' It transpires that octopamine is an invertebrate version of our dopamine, similarly involved in mediating pleasure.

Now, going back to language or communication and its role in social cohesion and reward for an individual, honey bees 'talk' to each other through the waggle dance, which they perform to communicate distance and direction to good food sources. Gene discovered that if you administer cocaine to bees it activates their octopamine system, intensifying their dance. You can train bees to sup nectar of marginal sugar quality, and the individuals that have been dosed with cocaine will still go back to the hive and rave about it. There seem to be some striking similarities between humans and bees here. It's as if a coked-up bee finds it hyper-pleasurable to spread the good news to its fellows. (Though I can't help wondering about the social implications since the news is not in reality as good as all that. Is there any comeback for the bee who has led its pals on a merry dance?) When we set aside this thought, the study neatly links the functioning of an individual's reward system to securing benefits for the community as a whole.

A second interesting finding from Gene's work is that within any population a small group of bees aren't social at all. They won't react if their queen dies, or if a new bee intrudes on their territory. Gene measured the gene expression in the mushroom bodies of those bees and found that about a thousand genes were differently expressed in them. He also discovered a statistically significant overlap between these genes and those that were associated with similar lack of social traits in some humans. 'We are not trying to make a glib comparison here,' Gene insisted. 'Bees are not little people. People are not big bees. But there are similarities between us at a molecular level that suggest that social brains, which have evolved at several places on different evolutionary branches, use similar substrates and are implicated in similar behaviours.' He speculated that perhaps these socially unresponsive bees, like their human correlates, function as a highly specific asset to the tribe. If an emotionally charged calamity strikes, for example, a calm, stable minority unaffected by perceived social effects will be capable of continuing as normal, improving the chances of future generations.

It's future generations – bringing them into existence and nurturing them, ensuring they have a functioning tribe to grow up in – who seem to be the key to the grand puzzle of love we've been investigating throughout this chapter. Earlier we explored how our appetites are shaped by things that took place long before we were even conceived. In this chapter we've seen that the act of sex and its derivatives – our emotions – seem to be all about the future.

Does that mean that as individuals we are, on some level, mere transmitters, biological machines living out our lives at one moment in time, acting with the interests of the species in (subconscious) mind? Emerging neuroscience research does seem to suggest this. It may feel challenging but also offers the chance to relinquish some of our more ego-driven attachments to certain events in our individual life stories. We may have reimagined the magical feeling of fantasy love as a ripple of chemical activity in the brain, but we've also seen that real long-term love is by no means impossible to sustain, and made the reassuring discovery that human beings, like other species, are pre-wired to seek out and give affection and care. This last point will have crucial relevance when we go on to consider the implications for society of the neuroscience of compassion.

CHAPTER FIVE

The Perceiving Brain

Our brain brings a predetermined package of capabilities and responses to every situation. It directs our interaction with our environment and is in turn moulded, through the formation of new neural pathways, by the thoughts and emotions that occur as a result of these interactions. In the previous chapter we looked at our brain's role in our intimate and emotional life. We saw how our social instincts are driven by features of our brain, and how these might play a part both in the personality of the individual and in generating collective relationships. In this chapter we will be taking things up a level of abstraction to look at how the feedback loop between the physical matter of our brain and the life experiences it processes takes place from moment to moment as well as day to day and year to year.

The brain is always operating on numerous different timescales. There's the one we're conscious of as we track through our various life stages, in which we learn to tie our own shoelaces, play the piano or interpret the mysterious behaviours of other people. Then there's the vast barrage of processing that takes place, much of it without our conscious awareness, in every second and millisecond of our waking life. Our brain must constantly engage in perceiving its surroundings and constructing a coherent model from them so that we can go about our lives. This is a staggeringly complex task, albeit one that we rarely notice unless or until it goes wrong. The brain interprets the signals it receives via the senses, and uses them to create a version of reality that we can function within.

We will be looking at this process in some detail because perception is the basis upon which all belief systems, simple and complex, are built, and our beliefs, though they may feel like some of the most idiosyncratic things about us, are as much subject to the biological constraints we are born with as any of the so-called lower cognitive functions we've examined already. Later we'll be thinking about the neuroscience of specific belief systems, including religion and politics, but for now, we're focusing on how our brain constructs our own personal reality, since this is essentially the platform on which the software of all beliefs is run. Everything you take to be true, from the apparently banal, such as 'I believe the sky is blue', to profound leaps of faith, such as 'I believe in God', depends on your brain's mechanisms for perceiving what is outside yourself, whether it's a physical object or somebody else's opinion, and processing that input to assign it a meaning and enact a response to it. Our ability to make decisions, collaborate, create, invent and hypothesise all flow from here. Consciousness, personality and life outcomes all depend, ultimately, on our brain's ability to construct a satisfactory version of reality.

Bespoke reality

The word 'version' is crucial because, as we will see, there is really no such thing as objective reality. I don't mean to suggest that the physical world does not exist, rather that every single person on this planet perceives it in a slightly different way. Everyone is living in their own bespoke 'reality', courtesy of their own particular brain's characteristics, its unique distortions, inherent filters and cognitive bias. Your perception of the world is not an accurate 'snapshot' but simply a subjective illusion. It is based on what you have seen before.

Take the example of 'The Dress', which became a viral sensation back in 2015 when a woman shopping for a dress to wear to her daughter's wedding took a photo of this soon-to-be-infamous item of clothing on her phone and sent it to her daughter for an opinion. They couldn't agree on what colour it was. One said the stripes were black and blue, the other said white and gold. Unable to believe that her mother couldn't see what she could, the prospective bride shared the photo on social media asking for opinions. Within a week it had attracted more than 10 million tweets as people vehemently disagreed over the colours of the dress.

This vigorously defended difference in perception is largely based on context, prior experience and expectation: if you are an early riser you are more likely to adapt your colour vision to the expectation of natural daylight and to see The Dress as white and gold. Night owls looking at the image under a yellow-tinged light are more likely to perceive it as blue and black striped. Our individual sense of reality is a construct, and the potential for differences in the reality that different individuals experience is vast. Your experience of each day is shaped by how you interpret the vast swathes of information constantly bombarding your brain from all of your senses, and all this information is processed through the lens of your unique prior understanding of the world.

At its most simplistic, this means that you tend to see what you expect to see in any given situation. This reinforces your evaluations and opinions, acting as a feedback loop to shape future perceptions so that your reality continues to fit with your view of the world. Since there are more than seven billion people on this planet, this means that there are more than seven billion different realities, each owned by the particular set of 100 trillion brain connections busily conjuring them up. Even more incredible is the realisation that your perception of the world, this non-real reality, instructs how you interact with it. It shapes the decisions you make, from the everyday matter of which sandwich to have for lunch to whether to go back to college and/or try for a baby. Everything about you is based on a unique hallucination forged by the interaction between the physical composition of your brain and your past experiences.

In this chapter we'll be looking at perception illusions, as well as the delusions suffered by people with schizophrenia, and the effects of psychedelics, in order to appreciate the vast amount of work that goes into the manufacture of what we blithely call 'reality'. The business of perception is fundamental to countless cognitive functions, and offers a prism for investigating the mechanics of how the brain produces consciousness and personality.

The faults in our filing

The first thing to grasp here is that even for those of us who are fortunate enough not to be suffering from a neurological disorder or mental-health condition, our perception of the world is not only highly individual because of our unique life experience but also inherently faulty because of species-wide flaws in the way the brain processes information. It may be routinely competent at stitching together a working understanding of its surroundings but it isn't always completely up to the job. It must take shortcuts to cope with the vast scale of the task. Those shortcuts lead to errors.

But why would our immense, sophisticated and powerful brain settle for providing us with an approximation of the world? If perception is indeed the platform on which so many other cognitive functions depend, surely it's worth getting it right. Why deal in illusions rather than accurate reality? Wouldn't that be less likely to result in potentially disastrous errors of judgement?

The answer is that the brain is busy. Very busy. And perception is just one of virtually countless tasks it's tackling simultaneously. To generate a working version of reality it has to transduce incoming signals from our ears, eyes, nose and other sensory organs to charged sodium and potassium ions, and pump them in and out of nerve cells. It must direct the resultant zip of electricity at speeds of up to 250 m.p.h. around our connectome, the most intricate and complicated circuit board imaginable.

This process is vastly energy-demanding. Given how hard it has to work, your brain, unsurprisingly, has evolved some tricks to make life easier. It would require even more energy to process every single detail and interpret all the signals the world emits. So it prioritises some information and ignores other bits. It uses prior experience to filter signals into either the energy-consuming processing pile or directly (and unconsciously) into the bin marked 'unimportant trivia'. This filing of information creates shortcuts in processing, which play a vital role in the rapid analysis necessary to form your ever-changing, up-to-date view of the world.

The problem is that the filing system is not infallible. Human beings are highly prone to perception glitches, flaws that are the unavoidable by-product of the mesmerising intricacy and complexity of our functional capacity. This is the case even in completely healthy people.

Some neat illusions demonstrate our brains' stubborn resistance to overriding their hard-wired perception shortcuts. Take a look at the two images of a hollow mask. The image on the right is, in fact, the back end of the mask, and the one on the left is the front. But even knowing this, you probably still perceive both images as a face, with the nose, eyes and lips all pointing outwards in a convex configuration, even though the shadows in the right-hand image are telling you that the opposite is the case: they are all pointing inwards.

 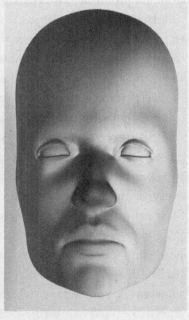

Our bias for arranging shape patterns into faces is virtually universal across all of humanity. We are, of course, extremely used to seeing faces everywhere we go in the world. This template of expectation is so strong, it has been built into our brains'

deep circuitry of perception over so many millennia that when we look at these images we ignore the cues the shadows are giving us and work on the assumption that the reverse of the mask is simply another face.

Your brain constantly employs prior experience to make assumptions about what it perceives. It's a crucial skill that has massively contributed to our survival, enabling us to make rapid inferences based on past occurrences and comprehend the world even as we are bombarded by its incoming signals.

The illusion above also helps us to begin to appreciate how we can each hold widely different views, since our individual sense of reality is based on a composite of our unique set of past experiences. Although we are exposed to many global experiences in life – we are surrounded by faces rather than hollow masks, for example – which facilitates some species-wide consensus around what we 'mean' when we refer to very common objects, such as trees and faces, what we term 'reality' is essentially an individual construction. Even within the consensus there is still scope for every one of us to perceive infinite unique versions. It is the combination of those unique versions that produces our highly individual perception of the world.

There is, as you probably (hopefully) won't be surprised to hear by now, a complex interplay at work between inherited brain circuitry and environment-shaped learning. To facilitate the task of constructing a serviceable version of reality, our brain physically changes its circuit board as it learns from the outside world. As it acquires knowledge it alters its connections between cells, allowing new pathways of thought. As we discovered earlier, our brains retain this flexibility and dynamism throughout our lifetime (admittedly to different degrees).

But it is this malleability that (counter-intuitively) constrains future information processing. Though your brain's neural pathways can and do alter as a result of the information they have processed, this alteration then becomes 'the new normal' and forms the basis for how your brain can perceive and process information in the future. On a neural level we have a tendency to see (or hear, or touch) what we expect to see, because our expectations of the world are nothing but the sum of our previous experiences.

Sensory overload and unreliable reality

So far we've been looking at perception glitches that affect all of us. These are the design flaws that evolution has yet to iron out. They are by no means trivial in their implications for what we understand about the nature of reality but most of us can and do live with them. We've probably all experienced the momentary disturbance of a spooky sensation – is that a face in the shadows or just the leaves of the trees? – but for the most part we're not even aware of these glitches, as our brain does such a good job of adjusting our sense of reality to cope with them.

Unfortunately, the brain's extraordinary ability to construct a coherent and stable illusion of reality sometimes goes disastrously wrong, as is the case with the 25 million or so people worldwide who have been diagnosed with schizophrenia. They may experience profoundly disordered perception (psychosis), with symptoms including delusions and hallucinations. The result is that they operate with a set of beliefs about the world that can be at odds with the consensus, a situation that is often destabilising and distressing for them and others.

I worked with a number of people who suffered from schizophrenia during my time at the psychiatric unit and their experiences affected me deeply. I was thinking about them, and about a dear family friend who was later diagnosed with it, when I wrote my PhD on perception and schizophrenia. Our friend spent weeks becoming increasingly agitated and terrified in his own home, particularly when he was sitting in the back living room. He was convinced that murderers were waiting for him in the garden, hiding among the shrubbery. Eventually he had a breakdown at his local supermarket and the police were called to help protect him and others.

Schizophrenia is a reference point in various discussions throughout this book, partly because it has been extensively studied so the data is generally robust, but also because it is a condition that continues to interest and move me. It is an umbrella term that covers a range of symptoms, but looking at its common features and neural architecture allows us to understand more about the brain's threshold for information-processing and the mechanisms that underpin perception. Essentially, it is a useful prism through which to consider how consciousness is generated, which makes it interesting to specialists in a range of disciplines, from biology to philosophy.

Returning to the hollow-mask illusion, people with schizophrenia are generally immune to its effects. They will see the image in its literal sense – as the back end of a hollow mask. (A word of reassurance here for anyone worried that their brain hasn't fallen for the visual trick: it doesn't mean you're schizophrenic if you don't see two faces. Psychiatric diagnosis is not that simple!) These individuals are collecting the same information as the rest of us with their eyes, but the 'top-down' process of interpreting and making assumptions is altered. Analysis of the brains of people diagnosed with schizophrenia has demonstrated that they have fewer connections in the circuits involved in learning, memory, reasoning, flexibility and higher cognitive control (the hippocampus and orbital prefrontal cortex). These areas also have a slower speed of electrical activity. Taken as a whole, this means that schizo-phrenics lack some of the hardware to filter information based on past experiences and to use this knowledge to help inform their perception of what they are currently experiencing.

As you can imagine, this makes the world a confusing place, with unfiltered information bombarding the brain via all the senses. People with schizophrenia are essentially exhibiting an extreme uncensored version of flaws in the circuitry that we all carry. As a result, their most fundamental beliefs about the world are very different from the species-wide consensus. Where we might unthinkingly assert that the sky is blue and the bushes in the garden contain nothing more threatening than next door's cat, a person suffering from schizophrenia is justified in thinking it highly probable that this is not the case. They are not inventing or elaborating so much as faithfully describing a world they perceive in radically different ways. And if they react with fear to what they perceive as a hostile place, perhaps that reflects the stress of being overloaded by the constant stream of signals that their brain lacks the hardware to process. The schizophrenic brain is one in which the interplay between physiological circuitry and life experience is catastrophically one-sided. Experience counts for little when the brain lacks the ability to learn from previous encounters and filter the important from the trivial.

Schizophrenia is largely a question of genetic inheritance, though the exact mechanisms that account for the disorder are complex. More than 180 genes have, as of 2018, been implicated in its development. Most play a role in laying down and main-taining deep neural circuits in the brain. A large cluster encodes proteins that specifically help with the brain's connectivity. These inherited genetic codings result in faulty proteins that impair connectivity in the mind, altering the way that a person learns, remembers and perceives the world, impairing the establishment of a trustworthy version of reality. It's estimated that approximately 80 per cent of cases of schizophrenia are the result of someone inheriting a mix of these risky genes combined with exposure to an environmental factor, such as infection during their mother's pregnancy, or drug use in later life (both of which could detrimentally affect the wiring of the brain's circuit boards).

Of course, the symptoms of psychosis do not necessarily arise from schizophrenia; neither are they necessarily linked to risky gene variants involved in laying down the foundations of brain circuitry. Some people experience psychosis as a result of their immune system going haywire and attacking the connectivity proteins in their brain cells, a recently discovered condition known as autoimmune encephalitis.

Professor Belinda Lennox from Oxford University, who has conducted research into this phenomenon, has found a way to treat such sufferers by filtering their blood to remove the elements of the immune system that are attacking the brain. It's early days for the trials on this treatment but it's estimated that approximately 5–10 per cent of first-episode psychotic patients, who may previously have been diagnosed as schizophrenic, could be helped in this way. The treatment effectively wipes out the radical perception glitches that trigger the patients' delusions and restores them to a less alienating and more workable version of reality, simply by correcting their immune system.

Alternative ways of perceiving

The experience of people suffering from schizophrenia is one extreme illustration of the fragility of any individual's most fundamental beliefs about the nature of the world around them. Taking psychedelics (such as LSD or psilocybin) can also induce perception of an altered reality, and these substances have long been of interest to psychologists and neuroscientists as a result. Given that they are illegal, it has been difficult to secure funding for studies investigating how they operate, particularly after the public uproar in the mid-seventies when it was discovered that the United States Central Intelligence Agency had surreptitiously tested LSD on unsuspecting human subjects, hoping to use it as a tool for mind control to manipulate beliefs. But nearly half a century later, there's been a resurgence of interest in these pharmacological agents. Thanks to the possibility of crowd-funding research, studies have recently been conducted in this area, trying to discover exactly how these mind-altering drugs work.

David Nutt is a professor of neuropsychopharmacology at Imperial College, London, and a leader in this field. He has spent his career investigating the effects of various legal and illegal drugs on the brain, and is an expert on addiction, anxiety disorders and sleep. He's also unafraid of controversy, having been dismissed from the British government's Advisory Council on the Misuse of Drugs in 2009 for claiming that alcohol and tobacco were both more dangerous for individuals than ecstasy, LSD and cannabis. He's genial, wears his expertise lightly and has a gift for explaining it. I wanted to ask him about his latest experiments in investigating the effect of LSD on the brain.

David and his team scanned the brains of volunteers, who were either given a 'micro-dose', a tiny quantity, of LSD, or a placebo. He was particularly interested in how being on LSD, even in very small quantities, affected activity levels in different brain regions. He observed that the drug switched on pretty much the whole brain. There was increased connectivity across all brain regions. People reported significant visual hallucinations, an inability to think analytically and a lack of interest in doing so, and a pleasurable feeling of the ego 'dissolving'. Essentially, the drug seems to loosen the brain's preference for working from the assumptions it has learned from prior experience. This profound effect on perception allows it radically to change its standard way of thinking. In some ways it could be said that the psychedelic induces a mild temporary version of schizophrenia. LSD can exacerbate hallucinations and delusions in those already diagnosed with schizophrenia, and has been implicated in its sudden onset in people already harbouring a genetic predisposition to it. But for healthy individuals taking mild doses, who do not have a propensity for schizophrenia, it may offer potential benefits that outlast the active experience of a trip.

What exactly is going on in the brain of a person on LSD? All psychedelics disrupt the top-down processing of visual information from the outside world. The hallucinations a person typically sees – distortion of shape and colour, cobwebs – are evidence of primary visual processing at work, without the usual filters being applied. The brain's visual cortex is suddenly connected to regions that it doesn't normally communicate with, allowing the person to conjure up sometimes staggeringly complex hallucinatory worlds, which are frequently filled with spiritual significance for them. As David puts it, 'The brain on LSD is being allowed to ditch *a priori* processing and do its own thing in a way it hasn't done since you were a baby.'

David is interested in using small amounts of LSD and other psychedelics, such as psilocybin (found in magic mushrooms), to treat people suffering from conditions such as depression, obsessive-compulsive disorder and post-traumatic stress disorder. These substances enable them literally to see the world anew, and to reconfigure their unhealthy beliefs. 'It's a bit like defragmenting a hard drive,' he told me. 'People report a strong psychological sense that they don't have to comply with negative thoughts. In fact they might not even have them any more. This is the great difference between the psychedelic LSD and the analgesic ketamine, for example [a drug David is also working on]. With ketamine, you can temporarily suppress negative thoughts but you can't change them.'

It's not yet clear exactly how LSD brings about this alteration in habitual thinking, though David believes that the drug's action on the brain's 2A subtype of serotonin 'is critical in allowing people to reset their thinking. It provides plasticity and opens the brain up to new learning.' More investigation will be required to establish the precise mechanisms at work, but in the meantime, preliminary clinical trials have established that psychedelics look promising as a treatment for post-traumatic stress disorder, addiction and depression.

The therapeutic use of psychedelics makes absolutely plain the link between perception and belief formation. A 'disordered' belief that creates negative impacts in somebody's life – if they are stuck in a habitual negative mode of thinking, for example, ruminating on past trauma, believing their life is worthless, obsessing or making negative associations and forming erroneous beliefs – can be shifted towards a healthier belief about their personal safety or competence by altering the circuitry of perception to bring about a different conception of reality. When reality is not an objective and immovable concept, what we believe about it, and how we decide to respond to and interact with it, becomes much more fluid.

Getting round the glitches: collective consciousness

If opening up the doors of perception doesn't appeal, there is a less risky, totally legal and easier way that we can mitigate the glitches in perception that can lead to disordered beliefs: get out and expose ourselves to a new experience or a new opinion – test our own construction of reality by comparing it with somebody else's. Professor Chris Frith from University College, London, has performed some groundbreaking studies in this area, bolstering the idea (of which we are probably intuitively aware) that when we discuss our subjective views with somebody else, we increase our chance of arriving at a more accurate representation of the world. The old adage that two heads are better than one appears to be true on a neurological level. Our ability to reflect on our perceptions and pass on our knowledge to others to affect *their* framework of reality, and vice versa, has the potential to help us arrive at a more nuanced understanding.

For this to work, though, we need to expose our worldview and opinions to challenges rather than simply have them confirmed, and as we've seen, on one level our brains are resistant to this. New information that requires us to re-evaluate our existing assumptions about how the world works is unwelcome, since it comes with an energy and attention cost. Our brains are very good at filtering out such challenges. This helps to explain why it can be so hard to change our own mind or somebody else's. We will be coming back to the neural basis for belief change in the next chapter.

But another competing mechanism comes into play to counterbalance the brain's innate conservatism in this regard: the drive to move, explore and seek out novelty. We seem to be hard-wired to enjoy (subject to the usual caveat of 'to a greater or lesser degree') meeting other people to share ideas and our perspective of the world. It helps us form a collective consciousness, which is really a way of referring to the mass of ideas that circulates in the world, feeding into and sustaining culture and investigation of all kinds. One of the great benefits of pooling of intellectual and creative resources is that it can help us to get round the bugs in our individual perception and belief-building systems.

As we discovered when we were looking at love and relationships, social interaction, communication and the reward pathway are all intertwined. Interaction with others can feel pleasurable on a biochemical level. This process drives us to build belief systems as a collective. The potential benefit, for individuals and cultures alike, is a more robust belief system that has adapted itself to the challenges of competing ideas, one that allows for more flexibility and creativity in the making of meaning.

Now, it's important to acknowledge that collectives are just as capable of sustaining disordered beliefs as individuals. If we only ever exchange ideas with people who see the world as we do and don't challenge us, our collective consciousness grows bloated until it's staggering under the weight of its internal contradictions. There have been too many tragic examples of disordered belief systems allied with state power leading to mass murder for us to feel complacent about the ability of groups to come up with good ideas. Discussion, even very sophisticated debate, will not inevitably lead to increased fitness of ideas unless it seeks out alternative points of view, listens to and reflects on them. The echo chamber of opinion is now more prevalent than ever in some circles, thanks to social media, and can supply us with the false sense that our version of reality is robust when in fact it is highly selective. But, even so, we can say that contact between people with differing opinions has brought about shifts in our collective values and cultures countless times throughout history. It is an ongoing process that will not be derailed, even if it cannot be relied upon to lead seamlessly to the ideas we would prefer to see flourishing in the world.

This innate pleasure in new ideas and novel experiences as a mechanism for mitigating our perception flaws is another driver of our propensity to move around our environment. As we saw in the chapter on the developing and ageing brain, movement helps boost the birth of new brain cells, and enhances brain health. Pre-wired into our reward systems are circuits that motivate us to explore our physical worlds, to interact with novel environments and people in order to keep this system freshly updated. As a species, and particularly in our younger years, we are inclined to roam about looking for novelty, with children and teenagers driven by this imperative for their rapidly developing brains. But, of course, learning continues throughout our lifetimes, with new pathways of communication being forged in the brain as we think. This flexibility, the phenomenon scientists call 'plasticity', is essential for our individual survival. It allows us rapidly to alter how we perceive and interact with the world. It provides an up-to-date map of reality as our world changes around us, which it does all the time, from the banal act of hopping on the bus to go to work to the more challenging arriving in a city we've never been to before. Wherever we are and whatever we're doing, our brain provides us with a constantly refreshed survival manual of our environment. It allows us to navigate our worlds.

Do some individuals have a brain type that supports a heightened urge to seek out new sensations? Some people seem intrinsically content to stick with their routine. You may be the kind of person who dislikes disruption to your cherished patterns of activity. Or you may be more impatient and restless, constantly craving new experiences and sensations, driven by the pleasures of the unexpected. Are these opposing dimensions of personality ingrained in each of us at birth? Or does our life experience cause them?

As always, the answer is 'a bit of both', but the evidence that these are highly heritable predispositions is mounting, with recent studies indicating that up to 60 per cent of a preference for novelty and sensation-seeking is down to genetics. In particular, the genes associated with these traits are related to, you've guessed it, the neurochemical dopamine system, which helps to prime the brain for reward. In essence, it seems that if you are born with the genetic complement for heightened dopamine functioning (specifically the D4 receptor subtype), then your brain is primed to seek out new sensations. If you also have the genetic package that means you get more dopamine in response to novelty, then, bingo, you're driven towards a life of reward-seeking explorations.

The propensity to explore different environments obviously has advantages for humans in evolutionary terms, and still holds benefits today. Some studies have demonstrated that individuals who are successful entrepreneurs or innovators tend to display high levels of novelty-seeking behaviour. Unfortunately, in today's environment such a genetic repertoire also carries a tendency towards less desirable activities, such as use of drugs of abuse that hijack our reward pathway.

There are life lessons to be drawn here. For those of us who suspect that we have a genetic predisposition for sensation-seeking, the secret to a happy, productive life is perhaps to fill it with adventures, stimulating new experiences and tales of the unexpected. There is, of course, nothing wrong in preferring to stick with routine and the known. That said, gently challenging oneself in a safe and secure way can be hugely beneficial. After all, as we discovered earlier, learning new skills, keeping active and taking on board other people's perspectives is important for long-term brain health.

There's no doubt that whether we're thrill-seekers or homebodies, movement is intrinsic to human behaviour. It enables us to interact with the world, express our emotions through gestures and speech, share ideas and reproduce. We are the extreme opposite of the humble sea squirt, a rudimentary animal that spends its juvenile days exploring the oceans, then settles down on a nice rock. The first thing it does after implanting on that rock is digest its own brain and nervous system. Then it sits there, eating organic bits that happen to float past with the current. It's a hermaphrodite, meaning it has the reproductive organs of both sexes, so it can reproduce without moving, and it doesn't appear to mind this rather sedentary, solitary existence. Our species, by contrast, is endlessly restless, driven constantly to update and refresh our ways of thinking during the course of our lifetime so that we evolve mentally as individuals and contribute our own (albeit partial) worldview to the collective consciousness of human thought.

Perception, bias and chauvinism

A great deal of restless novelty-seeking is observable in human behaviour, but a group of scientists dubbed the 'movement chauvinists' assert a more strident version of the movement hypothesis. For them, given that movement enables humanity to continue to thrive and evolve, to accumulate wisdom at an individual and social level, moving is effectively our sole purpose in life. Meeting new people and sharing ideas for the benefit of the species-wide consensus on reality, thereby fostering a more successful collective consciousness, is for them the 'meaning of life'. One might describe this as the neuroscientific answer to humanity's eternal quest to answer the question, 'What is the point of it all?' Movement chauvinism, like all belief systems, has evolved through countless iterations of contact between human beings, who share their glitch-prone and deluded versions of reality, their theories and imaginings, in pursuit of a more robust and more representative vision of the world. Like all belief systems it remains provisional, but to my mind it has an elegant service-ability about it.

It is, of course, useful to remain sceptical about even one's most cherished beliefs and to remember that they may well be built on flawed models of reality. When it comes to the vexed question of whether the brain is 'gendered' male or female, much of the research and our reactions to the results read like a cautionary tale for neuroscience. The processing capacity of our brains wants us to seek out patterns. Our perception of reality is built on this pattern-seeking model, and as a consequence it can be difficult to resist, even when the pattern we see leads us to categorise unique individuals.

All of us, however much we may consciously try to resist bias, slot people into pre-existing categories based on sex, gender, race, age and any number of other perceived identities. The problem is that this can result in a reinforcing and amplifying loop of behaviour, where individuals feel consciously or uncon-sciously pressured to conform to their 'type'. The resultant conflict confronts us with the question, is gender simply a social construct? The question is vast and nuanced and has been tackled by many specialists in different disciplines, but in neuroscientific terms, might gender stereotypes be the result of our flawed perception system, the result of our inherently biased, lazy brain? From everything that we've learned about how our perception of the world is built, it seems possible. But does this view stand up to scientific scrutiny?

First, despite decades of research into the biological basis of gender there are still huge holes in our understanding of how sex hormones affect brain development and function throughout life. It's also important to emphasise that, aside from the obvious effects of sex hormones on the development of external characteristics that differentiate females from males, other effects cannot be categorised as exclusive and binary: there is massive overlap between individuals across the genders. This is not surprising since, after the first wave of development of sexual characteristics that takes place in the womb, males and females produce both sex hormones (testosterone gets converted to oestradiol via the enzyme aromatase so males also produce the 'female' hormone while females also produce testosterone from their ovaries).

Second, interactions with others can shape sex-hormone concentrations, so if people identified as male are treated in a stereotypically 'male' way, their concentration of circling testosterone will increase, affecting their behaviour.

Third, and this is of crucial relevance, our learned assumptions about the world, based on prior perception and reality-building, can affect our judgement of ourselves, other people and our respective traits and abilities. There is a study in which participants were asked to note down their gender in a short demographic section before going on to rate their mathematical and verbal abilities. This seemingly innocuous act of consciously referring to gender on the form significantly skewed people's perception of their abilities, priming them with the stereotypical idea that men are better at maths while women have ingrained prowess with language. It also knocked confidence in the opposite direction, and all of this took place without participants being explicitly directed to think about either their own gender or any gender stereotypes. Other studies have investigated this type of effect. For example, subliminally priming women with stereotypically 'girly' words, such as doll, flower, earring, compared with 'male' priming, such as hammer, car, cigar, 'changed the women's lens of self-perception', as the authors of the study put it, so that the 'girly' priming resulted in women reporting they enjoyed a significant preference for arts-related over maths-based activities.

It's interesting to see how our desire to make classifications, however crude, can shape our perceptions. For centuries, as a species, we have been attempting to bend our beliefs towards identifying striking, immutable differences between individuals and to group people according to some external trait. Unpicking the plethora of biased studies and treading through decades' worth of faulty belief formations, particularly pertaining to gender, cannot be properly tackled within a single chapter. But for further information I heartily recommend Cordelia Fine's compelling books *Delusions of Gender* and *Testosterone Rex*.

In the next chapter we will examine in more detail the mechanisms by which our brains generate meaning and how they develop those meanings into specific belief systems, from religion to politics, conspiracy to scientific theories. For all its roots in the fallible business of perception, our species' ability to construct the most exquisitely florid and fiendish systems of belief is truly awe-inspiring.

CHAPTER SIX

The Believing Brain

What do you believe in? Ghosts? Manchester United FC? Innate human decency? Our beliefs are as varied as we are. They speak to the core of our sense of self and exert tremendous influence over our choices, decisions and opinions, steering us towards certain experiences and away from others. But, as we are discovering, these choices are not under our conscious control to anything like the extent we assume. We can only make a choice or form an opinion about something based on how we perceive it and, as we saw in the previous chapter, perception is flawed and intensely personal. Even something as abstract as what we believe, which we might consider the ultimate product of our conscious mind's workings, is to a large extent determined by brain operations that take place without our conscious awareness.

Those operations depend on a combination of species-wide biological constraints and our own idiosyncratic blend of genetic inheritance and cognitive biases. We may feel that our belief in God, say, has been derived from conscious processes: deliberation upon a spiritual encounter or intense intellectual engagement with theology and philosophy. But even the most sophisticated beliefs are the product of countless subconscious and innate brain mechanisms laid down before we were capable of rigorous analytical thought. This is not to deny that we spend a great deal of time thinking consciously and analytically about our lives and our beliefs, or to assert that such activity is without practical and intellectual value, but it is important to emphasise that all conscious thought is predicated on glitch-prone and biased perception and reality formation.

I take a belief to be anything that somebody accepts as being true about the nature of the world. Beliefs come in many shapes and sizes, from the trivial and the easily verified: 'I believe it will rain today', to highly abstract suppositions: 'I believe in God.' Taken together they form our own individual guidebook to reality, telling us not only what is factually correct but also what is 'natural' or 'right' and hence how to behave towards one another and the natural world. This means that our beliefs stem from our unique consciousness, a word I understand as referring to the ability to form a subjective view of the world. Viewed in this context, the practical value of beliefs becomes clearer. Belief formation is not a luxury among cognitive functions so much as an essential part of the toolkit for navigating the world. That's demonstrably true for individuals, and there is a corollary for groups. Nothing beats having a common set of beliefs as an aid to getting things done. Collective belief systems are the foundations for cultural, social and political projects of all kinds. And beliefs are not merely practical: they also seem to be good for us. Countless studies have demonstrated that believing in something – it almost appears not to matter what it is – maintains brain health and increases an individual's self-reported satisfaction with their life.

But, given that all beliefs are the product of our fallible brains, they are prone to glitches as well as capable of delivering glories. This can lead to problems at both an individual and a collective level. Somebody who is suffering from disordered perception and deluded beliefs as a result of schizophrenia is living with a terrifying absence of reliable meaning. Even a neurologically healthy individual can get trapped in their own version of reality, convinced that their belief that one race, gender or tribe of people is superior (for example) is self-evidently true.

Once beliefs take on a systemic and institutional character they become awe-inspiringly powerful, for good or bad. Belief systems have led to murderous intolerance via religious or political zealotry. Societies will wage wars to defend them. They have, of course, also enabled humanity's countless achievements in culture, science, technology and every other field of endeavour. And there is really no limit to what human beings will believe. Our beliefs may range from the banal to the baroque and back again, but what unites us is the extent to which we can believe very deeply in the rightness of our beliefs, in the face of all sorts of evidence to the contrary.

So where does this extraordinarily powerful phenomenon we call belief come from? How does our brain create the guide to reality that forms the basis for our personality and steers or even determines our outcomes?

Asking how we come to believe what we believe is not a new phenomenon. From antiquity onwards people have wondered how individuals can hold wildly conflicting perspectives, and whether it is possible to say that one belief is more valid than another. Most of us in the modern Western world probably go along with the suggestion that beliefs are to a great extent dependent on the family, culture and society into which we were born. These days, our beliefs are not necessarily the 'truth' handed down from some omnipotent and inaccessible force. We may believe fiercely in a cause or have a strong opinion on an issue, we may find it difficult when somebody else's opinion differs from our own, but in general terms at least, most of us accept that we as individuals do not have an absolute monopoly on the truth. We know that our beliefs, however sincere, are also relative.

They're a construct. So who or what constructed them? This is a question that neuroscience is increasingly seeking to answer.

The neuroscience of belief is a vast and fascinating subject that poses all sorts of challenges. More so than any other area we've tackled so far, it requires us to dig deep into how innate traits, genetic inheritance and lived experience interact to create an individual's behaviour. It is one of the fields in which the staggering advances in neuroscientific research are yielding the most awe-inspiring results. We can now more fully understand how and to what extent our beliefs are determined not by conscious intellectual effort but by the subconscious operations of deep-brain circuitry. Although what we believe is undeniably shaped by our experiences, with input from our family and society, it is also fundamentally derived from the mechanics of perception. Our beliefs are formed by, and simultaneously sum up, our unique sense of reality, and since this dictates how we interact with the world, the effect is a constant reinforcing of the beliefs that we acquired early in our lives. Long before we have an opinion about politics or football, we have beliefs about the nature of the world. If, for example, you form the belief in early infancy that the world is a trustworthy place in which, when you are distressed, a carer will come to your aid, that belief tends to be self-reinforcing. The opposite belief, that the world is indifferent or actively hostile, can be self-perpetuating, with sometimes tragic consequences for an individual's outcomes.

Belief: what is it good for?

Before we go on to look at the neuroscience of religious belief and political ideology, we'll consider the neurological mechanisms that underpin the creation of meaning in its most general sense.

Why do human beings so persistently seek theories to explain the world and ourselves? Where does the deep drive to analyse and interpret come from, and what function does it serve?

Psychology professor and founder of *Skeptic* magazine Michael Shermer suggested in his recent book, *The Believing Brain: From Ghosts and Gods to Politics and Conspiracies – How We Construct Beliefs and Reinforce Them as Truths*, that our ability to form beliefs is integral to human evolution. Just as we have seen that love is in some sense a by-product of the drive to reproduce, Shermer makes a compelling case for beliefs being a by-product of our brains' incorrigible pattern-seeking, a skill that served a clear evolutionary advantage. Identifying the pattern of a shadowy predator's face among the jungle's foliage, for example, and predicting that you might soon be lunch so best to leg it, allowed certain individuals to survive another day and potentially pass on this skill to offspring. We explored this concept in the chapter on perception, discovering that sometimes these self-protective mechanisms inbuilt in the brain can go awry, as is the case with schizophrenia.

We can think of our brain as a 'belief engine', constantly trying to extract meaning from the information that pours into it. It does this by categorising and cross-referencing all the sensory inputs it receives to generate patterns. The goal of all this largely subconscious work is to enable conscious cognition to help us to predict and plan for the future.

This is a staggering feat, but it doesn't always work flawlessly. The brain has a weakness for generalising from the particular. Typically, once somebody has had the same experience in the same context two or three times, they are happy to assert that this reflects 'reality'. We essentially model our current reality based on our prior experience, and this predictive process helps us to plan for the future. It is absolutely crucial for shaping our behaviours through what's called the 'direct-experience pathway'. For example, you encounter an apple; you eat it, it tastes good; therefore you expect the next apple you find to taste good. It does, so you repeat the experience a few times and end up with the belief that apples taste good. Perfectly rationally, you start to seek them out.

We are not the only animals who build such associations from the world to inform how we act in the future. It's a pretty fundamental skill, observed across species, that helps to foster survival. In fact, even beings without a brain can do it. The common garden pea, *Pisum sativum*, can learn by association, much like Pavlov's dog. Instead of associating a bell with food and salivating on hearing it ring, or spotting a shiny apple and hearing the stomach growl, the pea plant can be 'taught' to associate an air current with light so that when it is placed in a dark maze it grows towards a fan. The plant is choosing which direction to send its shoots based on knowledge it has acquired from experience. Given that even a pea plant is capable of formulating a rudimentary belief to explain how its world works, we might, as a species, need to re-evaluate what we think about consciousness. The more we learn about the neurobiology of other organisms, the greater the dent to our perception of human sovereignty over nature.

Beliefs as a by-product of perception

Back to the mechanisms of belief formation in humans. In addition to the direct-experience pathway there is also the social pathway, in which information is transferred from person to person. We spend a large proportion of our lives evaluating what people tell us and deciding whether or not to incorporate it into our worldview. (And, just to blow your mind a little more, plants seem to have mechanisms for learning via the social pathway as well. More on that later, when we consider some of the social implications of neuroscience.)

In humans, though, the social pathway is of fundamental importance. We have evolved with the ability consciously to reflect upon the world and tell stories about it, to communicate our individual beliefs through language. Language has long been regarded as the pinnacle of human cognitive skills, and its role in our ability to theorise and communicate is both fascinating and crucial.

As we saw at the beginning of this book, the basis for our unique perception of the world is formed during our early years when we're like sponges, soaking up information and experiences. Then, during adolescence, the heady combination of synaptic pruning and the desire for novel sensations lead to another crucial period for shaping our core beliefs about the world and the self. By our early twenties our brain has assembled an explanatory narrative that is the basis for our adult beliefs. The problem is, once the brain has constructed a belief about something, however partial or flawed, it prefers not to have to revise it. And, given the scale of the brain's eagerness to assign causal meaning to casual events, it's easy to see how quickly one could arrive at an erroneous conclusion ('White people are superior') from essentially random occurrences. The brain then becomes invested in these beliefs, reinforcing them by looking for supporting evidence while ignoring contradictory information. Future reality starts to mould itself around the belief.

This description of belief formation mirrors what we can observe of the physiological process of the formation of neural pathways. In both cases there is a self-reinforcing loop. As with the mechanics of perception, when it comes to belief formation the brain is wired to take processing shortcuts to save energy.

In this regard you may want to consider your brain inherently lazy. On a deep neurological level, our brains are invested in maintaining rather than changing our beliefs. The extra effort it would require consciously to change our minds about something, to lay down new neural tracts for this new conflicting thought, might simply not be worthwhile. This is especially true of beliefs we share with other people, the ones that form our social identity, such as family or religious beliefs. In these cases it's not just a question of adjusting information but of rearranging relationships. The stakes are high. The implications for the age-old question of whether a person can ever really change their worldview, and by extension fundamental aspects of their character, are fascinating. We'll be coming back to them.

Clever, but not unique

All the mental activity involved in information-processing, perception, generating consciousness, conjuring up our beliefs about the world and using language to communicate them to others is no mean feat, so perhaps it's no surprise that as a species we should be proud of our cognitive abilities. However, our attachment to the picture of ourselves as uniquely 'thinking' creatures means we have constructed the concept of human sovereignty over nature, seemingly blind to the cognitive feats of other organisms. It also means that we're over-invested in the idea that we are free agents possessed of unfettered free will. We desperately want to believe that humans are not simply (clever) animals or, to use a modern metaphor, machines. We observe that we are able to make decisions and act on them in the world and extrapolate from there to insist that we therefore have limitless personal agency. We even start to use the beliefs derived from our frequently faulty prediction machine to try to rationalise our decisions after the event, and attribute meaning to our environment and our own lives.

There's nothing wrong with this as such. We wouldn't have the lives, lifestyles or cultural riches that we do without our species-wide self-regard, and it's understandable that we're proud of those things. That said, a certain amount of scepticism towards our own good press might be a useful corrective to humanity's collective ego, given that when we see ourselves as superior we typically see nature as expendable. Besides, perhaps we don't deserve quite as much credit for our ideas as we think we do. Just as romantic love is to a certain extent the emotional recognition of deeply ingrained mechanisms driving us to reproduce or to foster support networks for our future, so our beliefs may be the conscious mind's extrapolation of the deep information-processing required to build a coherent version of external reality. We're clever, in other words, but consciousness is just one of the many things our brain-body system provides us with, and it's not even unique to humans.

In 2012 a group of neuroscientists attending a conference on Consciousness in Human and Non-Human Animals at the University of Cambridge released a statement that has become known as the Cambridge Declaration on Consciousness. It asserted 'unequivocally' that 'the weight of evidence indicates that humans are not unique in possessing the neurological substrates that generate consciousness. Non-human animals, including all mammals and birds, and many other creatures, including octopuses, also possess these neurological substrates.' It's not just the previously mentioned pea plant finding its way through a maze in order to photosynthesise: it's crows problem-solving their way through Aesop's fable in order to gain food, and ants cultivating their own crops long before humans discovered agricultural techniques for ourselves. They are all demonstrating consciousness: the ability to learn from past experiences and to act with their beliefs about their present reality and predictions about their future in mind.

If we reduce our emphasis on the part played by the conscious efforts of individuals in generating complex beliefs, we rely more heavily on evolution as the engine of this achievement. It is certainly a long way from recognising the pattern of that shadowy predator's face in the jungle to the elaborate thought systems that enable organised religion or party politics to function, but the evolutionary advantage of our brain's efforts to create meaning is so strong that it has indeed taken us all that way. This is true both for individuals and for complex social groups.

Donald Mackay was a physicist based in the Department of Communication and Neuroscience at Keele University during the seventies and eighties. He had a particular interest in the way that neuroscience could illuminate our understanding both of human consciousness and of Christian theology. Mackay formulated an influential argument about the purpose served by beliefs. He conceived of a belief as a conditional readiness to reckon, a way of preparing oneself to interact with something or someone. In this view, beliefs are not so much an incidental abstract notion as an invaluable system that prepares an individual to interact with life. All of us hold a vast number of beliefs about the nature of the world, and we rely on this network of meanings every time we make a decision. Being able to hypothesise about future possibilities from a basis of already established conventions enables quicker and more creative reactions.

Over millennia, human beings have come up with intricate and elaborate beliefs that serve a huge variety of purposes, from social bonding to the development of art, culture and technology. Without the ability to generate and articulate beliefs, it would have been impossible for philosophers and scientists down the ages to develop their theories. The intellectual movements and collaborative efforts that have led to everything from the concept of universal human rights to the eradication of countless diseases would never have evolved without our brains' incredible ability to analyse and narrate. There may be glitches in our individual and collective systems, but humanity's myriad achievements are their own testimony to the value and effectiveness of the production of meaning and its deployment.

There is certainly social utility in all this belief building, but it probably won't surprise you to know that the evolutionary conserved reward system is, as usual, playing its part in making such activities not only useful but also pleasurable. One of the answers to the question of why we have beliefs is the truism that without them we wouldn't have invented the wheel, boats, sanitation, novels, opera, contemporary dance or aseptic surgery techniques. But in addition to all these wonderful outputs, they can also provide something more intangible: a significant boost to the wellbeing and happiness of individuals and society as a whole. They make us proud and lend us purpose. Beliefs can be tremendously rewarding. Not always, of course. Ideology has done a great deal of harm to countless societies and one could argue that religious belief that engenders feelings of guilt and shame about sexuality, for example, is likely to have negative consequences for the wellbeing of the person who espouses it. Despite these caveats, as a category of brain activity, belief has overall been good for us.

How and why do we get so attached to our beliefs?

Dr Mario Beauregard, a cognitive neuroscientist from Montreal University, carried out a study that has become the textbook example of the link between an individual's conscious belief system (in this case a deeply held belief in Christianity) and their happiness levels. Dr Beauregard asked a group of Carmelite nuns to recall past mystical experiences in as much detail as possible, and scanned their brains as they did so. He wanted to observe which regions of the brain were involved with this activity. Numerous brain networks became activated, with variation in activity areas between individuals, presumably based on their specific memory, associations and emotional response. But the area that lit up, over and over again, was the reward system. It was pleasurable for the nuns to recall these spiritual encounters.

There have since been other brain-imaging studies linking religious beliefs with the reward pathway. Again, the nucleus accumbens fires up with activity, even during anticipation of spiritual experiences, suggesting a mechanism whereby exposure to doctrine can become intrinsically rewarding and motivating for individuals. Religious practice also commonly goes hand in hand with synchronous chanting or singing, which, as we saw earlier, can produce strong feelings of social cohesion. Add a dose of glorious architecture, evocative smells, resonating sounds and the sense of belonging to a community, all of which generate the pleasure response, and perhaps it is not surprising that the faithful associate their beliefs with heightened wellbeing, thereby providing the impetus to strengthen the belief further. In fact, religious people generally score more highly on self-reported scales of happiness and contentment than those who profess no such belief.

Interestingly, though, it might be that the key thing here is belief rather than religion. These results have been replicated in studies on people with other varieties of shared beliefs. Watching your football team win an important match at the stadium with your friends can have an exceptionally euphoric emotional effect, much more so than if you watched the game by yourself at home on television. This spike of reward presumably consolidates your commitment to the team so your belief is strengthened that this is the year they will surely win the championship, your loyalty cemented for another season.

We've now established that the human brain's impulse to seek out patterns, use them to create meaning and develop those meanings into sometimes elaborate belief systems is, to a large extent, an innate and universal drive with significant evolutionary advantages. Beliefs are useful from a practical point of view and strongly associated with feelings of increased wellbeing and social cohesion. We haven't yet addressed in detail the question of why and how we acquire specific beliefs, but there is a clue in those studies of religious belief and the reward system. As a fascinating study into the associations between political conviction and fear – at the other end of the emotional spectrum to reward – has demonstrated, beliefs are at least as much derived from our emotional responses as they are the product of intellectual processes.

One study set out to examine brain activity in self-defined conservative and liberal volunteers, who were exposed to perceived threats. It recorded activity in the amygdala region of the brain, which is involved in activating circuits that direct the body to prepare for fight or flight. It's now been discovered that higher levels of the stress hormone cortisol are produced when a threat is perceived, which decreases the potential for connectivity in areas of the brain involved in reasoning, learning, flexible thinking and future planning. This seems to be a sensible short-term survival tactic: the future is put aside to deal with the dangers of now, but it also effectively throws our 'hot' and 'cold' cognition out of whack. (Put simply, hot cognition is thinking coloured by emotion; cold cognition is information-processing and decision-making. The brain conducts a permanent balancing act between the two.)

Interestingly, analysis of brain scans from conservatives and liberals who professed strong convictions showed that conservatives typically have a more sensitive amygdala than liberals. In fact, both the anatomy and the size of the region are different. Connectivity between cells in a conservative person's amygdala appears to be much more elaborate, and the region takes up a larger volume of the brain. Taking these results together, it is possible to conceive that conservatives are more sensitive to perceiving threat, and act with immediate safeguarding in mind. In contrast a 'liberal brain' will show heightened activity in the insula, involved in 'theory of mind', which can be described, loosely, as the ability to perceive others as thinking entities. Those of the liberal persuasion are also more likely to have a larger and more reactive anterior cingulate cortex, a region involved in monitoring uncertainty and potential for conflict. It is conceivable that this endows a greater tolerance for the unknown and for complex social situations.

You may be thinking that this sounds like delicious validation for smug liberals everywhere: the conservative brain limited by fear, the liberal brain replete with great capacity for creative collaboration. We should certainly be wary of leaping too quickly from levels of brain activity to political opinion without considering the complexity of belief formation in the middle. It's not that babies are born liberal or conservative, more that an individual's brain (a sensitive amygdala or a sensitive insula) may prime them to build a view of the world as a frightening place or a welcoming one. As we've seen, such foundational beliefs, established in infancy, may be overlaid with further beliefs about, say, the threats posed by people from beyond one's own social group who move to one's home town, or the dangers posed by technology, the threat of radical Islam, right-wing evangelical Christianity or any number of other potential dangers that are themselves constructs of a system of beliefs. It's a complex web of belief formation over the course of a lifetime that may one day yield the self-applied label of a liberal or a conservative.

We're right to be cautious, but if you're tempted to dismiss the findings, bear in mind that the US researchers who conducted these studies claim they could employ such brain scans to predict an individual's political bent, as either Republican or Democratic, with high sensitivity and accuracy. So perhaps it's fair to say that the results are strongly suggestive but the value judgement about whether the study demonstrates that one or other mindset – liberal or conservative – is 'better' is of course a matter of opinion. Your interpretation of what these results tell us about our world is highly contingent on your pre-existing opinions about everything from the relative merits of political ideologies to the efficacy of neuroscience, not to mention the relative size of your own amygdala and insula.

My own beliefs incline me to the interpretation that it's important that both brain types exist in our society. Perhaps the more conservative type helps to protect the individuals of the present while the liberal brain helps foster the success of future generations.

Can we ever change our minds?

The truly intriguing thing for me is to wonder whether political views can be 'reverse-engineered' by the application of this knowledge. When we expose ourselves to twenty-four-hour news cycles and scroll through endless feeds on social media, we are bombarding ourselves with what feel to the brain like danger warnings. Some small studies have indicated that it is possible to change a person's emotional state by altering their Facebook feed. This has an effect on cognition, altering the fine-tuning of perception and decision-making to put emphasis on the habitual, automatic, defensive system that has evolved to help protect us, at the expense of our collaborative, empathetic and innovative problem-solving potential. A more 'conservative' choice or decision may be the result.

The implications for our own sense of agency are profound. Much of what we regard as our conscious intellectual activity and its outputs – our opinions and beliefs – is shaped by emotional response driven by deep-brain functioning. The social and political implications are potentially alarming, especially in the wake of scandals over Facebook's handling of personal data and its use by campaigning and commercial companies. We'll be coming back to this issue later.

The possibility of engineering an 'ideology switch' raises the question of the extent to which we typically change our beliefs. As we saw at the beginning of the chapter, our brain's drive to expend as little energy as possible makes it, in some ways, inherently conservative when it comes to perception and the generation of meaning. But, of course, people do change their opinions, as the result of a dramatic event or through the slow accrual of lived experience. The saying, sometimes attributed to Winston Churchill, that 'If a man is not a socialist at twenty he has no heart but if he is not a conservative at forty he has no brain' encapsulates a change of opinion that many of us have probably witnessed, if not lived.

To find out more I spoke with Jonas Kaplan, professor of psychology at the University of Southern California's Brain and Creativity Institute. He researches the neural mechanisms of belief and has conducted some neat research that examines what happens in the brain as a person attempts to maintain core beliefs in the face of counter-evidence. His group studied self-defined liberals (easier to recruit than conservatives, apparently, at least in Southern California) and asked them to consider the evidence that Thomas Edison did not in fact invent the light bulb and that multivitamins are really not very good for you. Alongside these relatively innocuous challenges they were also presented with evidence to undermine their beliefs in the validity and usefulness of increasing taxation on the wealthy, tightening gun control and increasing access to abortion.

The volunteers were typically open-minded about the matter of Edison or multivitamins but extremely resistant to challenges to their political beliefs. As Jonas put it, 'Being left-wing doesn't correlate with open-mindedness. Our volunteers were very committed to their political values. They said things like, "How could I explain it to my friends if I changed my mind about this?" If being a left-wing person is crucial to your self-identity, you are highly resistant to change.' So although ideological leaning correlates with a particular brain type, people on either side of the political spectrum are resistant to altering their core identity beliefs once established.

So what exactly is happening in the brain when a core belief is challenged? Jonas and his team measured levels of activity in different brain regions to answer this question. 'Initially the brain shifts from using external cognition networks to internal cognition networks,' Jonas told me. 'This is what we would expect to see when people disengage from whatever is in front of them and shift to scanning memories or thinking about themselves.'

Essentially, when core beliefs are challenged people search their mental catalogue for counter-arguments to the evidence being put to them. They will attempt to slot the new information into their existing view. If that can't be accomplished they will try to disregard it to reaffirm their existing cognitive model. Jonas spotted surges of activity in the amygdala and the insula as this evaluative stage was carried out, suggesting that people's emotional responses to the new information were crucial in their decision-making about it. 'There is a lot of recoil from any information that challenges our belief systems,' Jonas told me. 'It is a danger to our core self, so the brain's self-protective system kicks in, regardless of whether you have the brain profile of a liberal or a conservative. Those people with higher levels of activity in the amygdala system were less likely to concede any ground.'

Jonas's research may be useful in the context of fostering flexibility of opinion. I asked him whether his work inclined him to believe that people could learn how to update their views on the world. Was there a way to present the mounting and compelling research on climate change to help even the most resistant to break out of the fortress of core-identity beliefs that their mind has constructed? Jonas is working on exactly this concept, attempting to train people to regulate their emotions and see whether it enables them to change their minds. He's using a technique called emotional reappraisal, in which he shows volunteers pictures they initially find disgusting. They are encouraged to think about the picture in a different way, to disengage from automatic responses and automatic inter-pretations. His preliminary results look promising and he believes that fostering a fearless curiosity for life rather than a knee-jerk self-protective response could have far-reaching consequences for individual creativity and wellbeing, innovation, entrepre-neurialism and even collective action and outlook. Although the initial results are not yet in, it will be interesting to see how this field of research develops and how we can use the results to create a healthy balance between individual self-protectionism and survival and a healthy enthusiasm and curiosity, especially in an age where much seems to be rapidly changing. We'll be asking in a later chapter whether the principle of training the brain for maximum flexibility could be incorporated into education programmes.

Belief change as therapy

Emotion plays a big part in belief formation and change but it's certainly not the only relevant framework. Recent research has investigated the impacts of physical movement and rest respectively on the way we develop and alter beliefs about our environment and our role within it. The implications for boosting mental wellbeing are profound.

Numerous studies have linked physical exercise to the functions of the reward system. We seem to want to move, and movement seems to play a part in maintaining brain plasticity, with all the attendant benefits for brain health. Meditation, meanwhile, which is almost the opposite activity, has long been associated with good mental health and is a pillar of spiritual practice in Buddhism and Christianity. Devotees claim it induces clarity of thinking and a less ego-driven attention to the world and the self. Now, thanks to imaging technology that allows us to observe which areas of networks within the brain are recruiting oxygen to power metabolic activity (functional magnetic resonance imaging, or fMRI scanning), it is possible to peer inside the meditating brain and investigate these assertions. A number of studies with long-term practising Buddhist monks have revealed that meditation engages clusters of networks within the brain, including the caudate region (thought to have a role in focusing attention), the medial prefrontal cortex (involved in self-awareness) and crucially the hippocampus (learning and memory). It has been suggested that over time meditation helps to support the process of neurogenesis by nourishing the newly born cells so that they form fully functional connections and circuits in the brain. Meditation also seems to promote the production of protective fat around these cells and to reduce the damaging effects of the stress hormone cortisol, helping connections to flourish.

Other studies have specifically focused on different levels of electrical activity in the meditating brain and have revealed that it is in some ways similar to that of a sleeping brain. Low-frequency electrical waves, typical of a sleeping brain, dominate during meditation but in a unique combination with electrical oscillations of a higher speed typically associated with cognitive work. Meditating seems to emulate some of the restorative and nourishing aspects of sleep but also pairs it with relaxed and creative thought, thereby helping to produce a state of memory-consolidating relaxed alertness that's highly conducive to good mental health and improved cognitive functioning.

It therefore seems that both movement and space for reflection and rest are required to boost brain health and enable people to maintain a flexible and open mind about themselves and their situation. In fact, despite the movement chauvinists' rather one-note insistence that their belief system explains the riddle of human purpose, movement's opposite state – stillness, a central tenet of religious belief systems for thousands of years – appears to be equally vital to individual wellbeing and the development of the collective consciousness. The two states have been shown to be deeply complementary on a neurological level.

In the light of these findings a new clinical intervention for depression has been developed, called mental and physical (MAP) training. One way to describe depression, like other mental-health conditions, is as a set of unhelpful beliefs about the world. I am not suggesting that the reasons for people's depression are not real, or that they can be cured simply by positive reframing. But if somebody believes, for example, that they have no value, that the world is better off without them, then enabling the person to alter the underlying brain mechanisms and therefore prevent these negative beliefs from persisting can be hugely beneficial.

MAP training is the highly practical attempt to take what neuroscience has demonstrated about deep-brain functioning in relation to belief formation and apply it in a therapeutic context. It combines set times for focused-attention meditation with a period of aerobic exercise, such as running, for thirty minutes. Preliminary studies show that this regime appears to be working: a group of young mothers who had recently been made homeless and suffered traumas had improved wellbeing scores after MAP training. It has also been shown to help those diagnosed with major depressive disorder and even enhance wellbeing scores for individuals presenting with no diagnosis but reporting general happiness already. Admittedly we can't yet say for sure whether the number of neurons and circuits in the brain increases as a result of MAP training. But these preliminary results illustrate how neuroscience research might be successfully translated into novel clinical interventions to benefit individuals' health and happiness. The hope is that people will be able to use MAP training to bring about healthy belief change in themselves.

I'm rather soothed that the last few pages have taken us away from the dystopian nightmare of 'ideology switches' that hijack the brain's fear responses, or mind-control by corporate or political interest groups, towards a much more empowering model for changing our minds and the views of others through the art of self-reflection and exercise.

Belief, fate and free will

Throughout this chapter we've been investigating the way that what we believe about the world and our self can have a profound effect on our experiences and outcomes. We've dug deep into the question of how our beliefs are formed and whether they can be altered, at a conscious or subconscious level, by ourselves or by others. The complexity of belief formation should not be underestimated since it relies on everything from species-wide perception flaws to the intricacies of multiple layers of our unique experience, and filters it all through the specifics of our genetically determined brain circuitry. This hugely complex system supports a myriad of self-referential convictions about belief, consciousness and the extent to which we are autonomous beings. The elephant in the room that we have been circling is the central question of the book. Are we subjects of biological fate or agents of free will? Do we really have the capacity for freedom of choice or are the decisions we make on a daily basis actually inevitable computations? Is free will simply an illusion?

Way back in 1985, neuroscientist Benjamin Libet devised an experiment that attempted to determine whether the conscious decision to move is made before or after our brain initiates the cue to instigate the physical movement. He asked subjects to flex their wrists repeatedly at times of their own choosing. He noted the moment the movement occurred and the activity of the brain's motor cortex, and compared those data with the time the individual reported consciously deciding to move their wrist. Precise times for the wrist movements were obtained by picking up the electrical activity of the muscles. Similarly, electrodes were placed on the scalp to record electrical activity in the motor cortex with high sensitivity. Libet discovered that the subconscious instruction from the motor cortex came first; the conscious decision to move came 350 milliseconds later. Then there was a 200-millisecond delay before the actual movement occurred. In effect, conscious awareness took place after the brain directed the action.

There are obvious limitations to this experiment. The time it takes the subject to focus their eyes and read the precise time on the clock may introduce errors, while an individual's reporting of the sensation of decision-making is also subjective. Most fundamentally, the experiment took place within a laboratory-constructed paradigm and examined a very simple one-step decision. That said, the experiment has been repeated and refined many times subsequently with similar results. But how does it relate to the plethora of complex and nuanced decisions we all make on an everyday basis?

Interpretations have varied widely. Some people have suggested that the time lag between your brain preparing to initiate movement and your conscious awareness of deciding to move provides a moment in which you could, theoretically, veto that action. Perhaps, extrapolating these findings into the real world, it's this inbuilt pause button in our minds that provides us with the opportunity to exercise our agency. That's a nice idea until you consider that it's now well established that impulse control and willpower increasingly appears to be determined by a combination of genetic predisposition and early learning as any other trait. Some people are predisposed towards taking advantage of the pause button and others simply aren't.

Given his work on belief change, I wanted to ask Jonas Kaplan for his thoughts on the concept of free will. His response was unequivocal. 'I don't believe in free will. The universe is deterministic. We are not the author of our own actions because everything is caused by something prior.' He did add this caveat, though: 'But decisions are partially controlled by our emotional state, and since most people find it depressing to believe that they have little or no free will, there is a lot of value in believing in it.'

It wasn't the first time (or the last) that I've had this view put to me by people working in cognitive sciences from a number of different perspectives. A slew of recent research suggests that eroding an individual's belief in free will fosters self-centred and impulsive behaviours. If people think their actions are largely predetermined they tend to decide that nothing they do really matters, and set about following their desires at the expense of social rules. Our individual belief in free will may well be an illusion but it's probably a necessary one for a smooth-running society (though also, potentially, for a smooth-running life).

Practising open-mindedness

To explore the need to believe in our own agency from a rather different perspective from my own, I spoke to eminent theologian Lord Williams, former Archbishop of Canterbury. Neuroscience is increasingly suggesting that our individual consciousness, our subjective view of the world and the beliefs we form about it, is just one of many things that our very clever brains produce through their electrochemical circuitry. I wondered whether, if Rowan does believe in free will, he considers that it derives from the physical constructs of our brains or from some other source. Is it of divine origin? Perhaps a soul that co-exists with consciousness is instrumental in human agency.

'I do believe in free will,' he told me, 'but I don't think there's a contradiction between saying, "This is how it is produced in the brain," and saying, "And this is how it actually works, in ways that are not completely predictable." To my mind it cannot be the case that every decision is predetermined. If it's possible for you to predict what I'm going to say over the next five minutes, it's also possible for me to say, "No, I'm not." So the exchange of information via language impacts on the deterministic framework.'

Perhaps, then, determinism holds on an individual level but adding another person's presence disrupts it. The number of variables that a conversation could introduce is, of course, theoretically infinite, making it harder to maintain an absolutely tight deterministic model to account for what somebody may say or do. (That said, there have been a number of fascinating experiments that show that, in practice, most of the things we say are so constrained by predictable variables [our relationship with the person, our background, the emotional context of the exchange, social expectations, the rules of the language, etc.] that a staggering amount of prediction *is* possible. We could say anything at all – 'Cucumbers make wonderful pets' – but we stick with what we know: 'It's very cold out.')

Rowan's reading of advances in the natural sciences has led him to think deeply about the matter of complexity and how it plays out. 'There seems to be a pattern in the way organisms evolve which at a certain point generates the capacity to image the self and imagine what isn't present. Once consciousness has developed, it seems to feed back into the organic world in a way that makes that world not entirely mechanistically predictable. The way it does that is via the thing we call agency, and it disrupts the system, so that I am not simply a mechanism.'

Rowan holds the view that language, and specifically linguistic interactions between people, are the tools by which agency exerts itself. I asked him whether he believed that he had been born with a predilection to imagine, to reflect, question and then to communicate his views and discuss them with others. Did his past experiences not inform the questions that he would ask and the answers he would give? 'Inform, yes,' he said. 'But there's a difference between informing and determining. My past is a factor. So are my predispositions. But is any of it so conclusively deterministic that there is no alternative way of acting in the future?'

Rowan is correct to say that, currently, biology has no way of properly answering that question. We cannot yet assert confidently that our past experiences and our neurobiological hardware predict rather than merely inform our future. It's also important to return to the concept of perception, and to consider that our recall of any event, any conversation, will necessarily be utterly idiosyncratic. Our interpretation of the words that were said and the emotional connections made to them will be shaped by our past experiences.

I find it possible to envisage a scenario in which even very complex human behaviours are largely explicable in biologically deterministic terms – after all, we've added genes, hormones and epigenetics over the last fifty or so years, with more advances coming thick and fast – but I concede that we will probably never know *everything* about causation of behaviour. Perhaps that's a good thing. After all, as Jonas suggested, free will may be an illusion, but it's probably a necessary one. Determinism may be neat in a theoretical context but, as Rowan pointed out, it doesn't work to translate rudimentary findings from a laboratory setting into the real world and conclude that doing so dispenses with the need to engage in the process of self-interrogation or picture another person's point of view. As Rowan put it succinctly, 'That doesn't seem to me a recipe for humane interaction.'

I asked him whether he thought that he might be hard-wired

to try to stop us all thinking we're machines. Laughing, he said, 'I'm probably hard-wired to want to make a difference, to help people be more reflective, to understand the variety of factors that flow into our humanity. I'm motivated to prevent us from thinking that our circumstances are wholly beyond our control or at least our intervention, because that can lead to a depressing static effect. In simple terms I'd like people to believe they can make a difference. I'm extremely wary of any system that suggests otherwise. I know you can say immediately, "That's your conditioning or predisposition." But I come back to the basic fact that once an idea is out there, even a statement of determinism, it becomes discussible.'

I found it hugely inspiring to discuss my beliefs in neuroscience and biological determinism with Rowan, who comes at the subject from an alternative perspective. It was also reassuring that there was overlap between new findings from neuroscience and the traditional studies of theology and philosophy. Rowan's emphasis on the mutability of all ideas brought me back to Jonas's experiments on the neuroscience of belief change. It's uncomfortable to change habits, never more so than when those habits are the mental ones we call beliefs. Rowan's personal belief in the vital importance of reflection, discussion and hopefulness left me feeling . . . well, hopeful. Perhaps we all need to be a little more like him, actively practising flexibility of thinking, compassion and curiosity.

CHAPTER SEVEN

The Predictable Brain

In the previous chapter we considered the highly abstract question of whether our cherished notion of being a free agent can survive neuroscience's investigations. I challenged myself to subject my own belief in biological determinism to a powerful different perspective when I talked to Rowan Williams about his belief in human agency. I came away from that encounter inspired by the dynamic effect of practising flexibility of mind. I remain convinced that we are on a path to ever greater neurobiological understanding of human behaviour but that this knowledge must be positioned in a real-world setting. If we do not achieve this, we risk falling into a less nuanced understanding of humanity and sleepwalking into ethical swamps.

In this chapter and the next we turn away from the abstract questions of how individual behaviour and the sense of self are generated in the brain. It's time to consider what we do, on a practical level, with all the new knowledge being generated by neuroscience. Medicine is our starting point, and the focus of this chapter, because it's the field in which neuroscience has already had significant practical impact, and where the concept of biologically determined outcomes is clearest. Medical researchers and practitioners are already asking the complex ethical questions that we will all increasingly have to wrestle with. As we discover more about how neurobiology shapes health outcomes we will be able to predict individuals' medical futures. But do we want to? Does it help to know that we are susceptible to Alzheimer's, Parkinson's disease or brain tumours, say, or is that knowledge likely to blight our prospects?

The question goes to the heart of how we feel about the place of autonomy in our lives. The answer, of course, depends on many things. As I discovered for myself when I found out I was a carrier for haemochromatosis, it's complicated even in relatively benign scenarios. Speaking to people who have had to decide whether or not to take the diagnostic test for Huntington's disease showed me that knowledge is not necessarily empowering if it confirms an unavoidable fate. It's one thing to know you're at risk of developing Alzheimer's when there might be things you can do to help mitigate that risk, and quite another to be told you're certain to develop a devastating disease for which there is little treatment and no cure. I found it humbling to recognise that in some (admittedly extreme) cases, ignorance may well be preferable to knowledge of one's destiny.

Before we consider the unusual and troubling case of Huntington's, though, we'll look at the breakthroughs being made in identifying biomarkers for the development of other mental health and neurological conditions, where there are significant grounds for optimism.

Predicting the future in order to change it

A biomarker is simply a measurable indicator predicting a biological state or condition. The presence of an antibody in the blood cells, for example, is a biomarker for an infection. Specific mutations in BRCA1 or BRCA2 genes are genomic biomarkers for susceptibility to breast cancer. Thanks to neuroscientific developments, biomarkers are now being identified that can predict, with increasingly high sensitivity and selectivity, whether a person may develop a specific mental-health-related condition, if they are prone to particular behaviours and how they are likely to respond to particular treatments. Conditions that were previously viewed with superstition and mystery are beginning to yield their secrets, and as diagnosis becomes more sensitive, treatment is on the brink of becoming more personalised and more effective. There are now reliable diagnostic tests that can predict whether a person will develop Alzheimer's up to thirty years before symptoms begin to appear. New tests predicting risk for developing Parkinson's and drug-resistant depression are being offered and it is increasingly possible to use biomarkers to personalise treatment for individuals with psychosis.

The UK government has clear hopes to build on these findings by embedding personalised medicine in NHS healthcare. In 2012 they spearheaded a multimillion-pound research investment called the 100,000 Genomes Project, sequencing genomes from people with cancer and other diseases, including anorexia and schizophrenia, to identify a greater array of biomarkers to help build the medical toolkit. Since a better understanding of the mechanisms underlying a condition and early diagnosis both improve treatment outcomes, the ultimate aim is to improve quality of life for millions of people around the world.

It is entirely justified to be excited at the prospect of such developments. When it comes to medical breakthroughs, the last century has seen advances at a remarkably progressive rate. Think of developments with keyhole surgery, organ transplant,

IVF, immunotherapy as targeted treatment for cancer, to name but a few. But this also feels like a good moment to remind ourselves of the dangers of neurohype. At the moment it is still much easier to predict and treat health problems that originate from other organs within the body. The brain has proved much more elusive in revealing its magic and has only recently started shedding its layers of complexity. Reliable prediction of the emergence of behaviours or personality traits in an individual is therefore a big ask. As I've emphasised in these pages, neurobiology is highly complex and it's not helpful to suggest that a single gene (or a single brain region or indeed a single anything at all) is responsible for any aspect of human behaviour.

Even with that enormous caveat still echoing in our heads, I do believe that in the future it will be possible to predict a great deal about brain health, temperament, skill sets, probable life outcomes and individual risks. Breakthroughs are constantly being made in the mapping of the human brain. As we saw earlier, we are now able to observe the connectome taking shape as a baby develops in the womb and study the impact on that brain circuitry once the child has emerged into the world, visualising how the connections grow and alter as the infant interacts with its environment.

The scientific panorama is changing fast as huge genome-mapping projects reach fruition and massive data sets are released. That data is coming online as I write, and initial studies are emerging from week to week. Genes linked to positive outcomes such as prosperity and more desirable character traits, including intelligence, creativity and strong willpower, are being uncovered. Some are also linked to other biological processes, including living longer.

Even the genes involved in the age somebody might lose their virginity have recently been identified, with genetics thought to contribute a hefty 25 per cent. They're the genes that direct onset of puberty, and are linked to risk-taking behaviour, impulsivity and sensation-seeking: the full adolescent package.

Commercial genetic-screening companies are leaping on all these findings, striking up collaborations with academic groups, rushing to be the first to come up with genetic tests for intelligence or creativity, while pharmaceutical companies are already striking deals for the rights to access this data. It's now possible to have your entire genome sequenced for a few hundred pounds and in a matter of hours. You can opt for a report focused on ancestry or a range of health issues, including genetic susceptibility to conditions such as Alzheimer's and Parkinson's, and being a carrier for the gene variants implicated in forty diverse conditions, ranging from autism spectrum disorder and cystic fibrosis to hereditary hearing loss.

Many of the people I spoke to while I was researching this book said they had their doubts about the usefulness of commercial genome testing. Ironically, the really helpful information it may provide may not be made available, since there is a legal requirement to offer genetic counselling alongside information about susceptibility to certain more serious illnesses. But there's no doubt that the market for genome-sequencing is only likely to grow. These advances, in combination with the current connectome revolution, will help us to understand more clearly how genes, brain circuit and environment interact. In short we are entering the era where teasing apart nature from nurture becomes possible. But in that profit-driven charge there will be lapses in credibility and a great deal of misinformation for sale. Neurohype writ large.

What are the risks in knowing the future?

In the short term, the effects on how individuals perceive themselves and take life decisions could be huge, and not necessarily positive. If you are able to take a commercial genetic test that tells you that you are at risk for anxiety disorders, let's say, you may draw all sorts of conclusions about what that means and how you should alter your behaviour. Will that information only exacerbate an already underlying susceptibility?

My opinion is that, although there will certainly be some pseudoscientific grabs of the new knowledge, there will also be more and more fascinating and credible studies, and applications that take a robust approach to predicting our future. At some point within the next decade or so we may be able to determine not only how healthy we will be over our lifetime but how happy, successful and wealthy. The question is (as it was for health outcomes), do we really want to know? Would it be empowering or would it lead to disillusion and premature disappointment?

A further complication arises if we ask whether you'll even be free to take this decision for yourself. You may decide you'd rather not know anything about your innate susceptibilities and opt to get on with your life as best you can, but it may not always be within your power to refuse that knowledge. Imagine that a healthcare provider requires genome testing before they carry out treatment. There might be benign grounds for this (to determine which treatments would be most effective for you) but unless we develop rigorous guidelines around genetic privacy and healthcare provision, more disturbing scenarios are possible. In the UK, healthcare remains free at the point of use for all residents, but that model is under pressure. We may move towards a model based on private health insurance, which is common in other parts of the world, and in that context it is possible to imagine that an insurer would insist on a test as a prerequisite for taking out a policy. They would also doubtless increase your premium or even refuse you if the results suggested you were heading for poor health.

There are already serious ethical questions to ask about using genetic profiling in the context of IVF or, potentially, as a routine part of prenatal screening. Should parents be entitled to decode their child's destiny? Given that currently the majority of UK women carrying a foetus with Down's Syndrome now terminate the pregnancy, having established this precedent, we need to ask which traits of the human condition we want to screen out through selective implantation of embryos and which we wish to retain in our population. How far, as a society, will we allow screening and the creation of 'designer babies' to go? These questions are particularly pertinent given recent advances in technology.

New test methods allow genetic testing just five weeks into pregnancy. The mother's blood is sampled, picking up foetal cells in her circulation that can then be isolated and screened. This is much less invasive than the traditional diagnostic prenatal genetic tests obtained via amniocentesis or chorionic villus sampling, both of which risk inducing miscarriage.

The development of such non-invasive prenatal testing (NIPT) does, however, come with increasing ethical considerations, which the Nuffield Council of Bioethics, one of the world's most influential bodies for setting standards of acceptability in the medical field, have recently deliberated on. These include the potential to undermine the future person's ability to make their own choices about accessing and allowing others access to information that relates to their health, abilities, personality or physical attributes. It may also encourage discrimination against people with certain genetic features, such as their sex, or contribute to damaging perceptions of what constitutes a 'normal' or 'healthy' baby. In light of these concerns the Nuffield Council advocates that such genomic screening should not be used, by either the NHS or in the private market, to screen babies for diseases that may emerge later in their lives. Screening should be done only for conditions that are 'serious but treatable', and only in those cases where there is evidence that genome screening will 'reduce ill-health or death'. But where, exactly, is the line drawn?

The issue is becoming more complicated by the day as technological breakthroughs allow for new applications in not just genetic screening but also gene editing. In the summer of 2018 the Nuffield Council, in a move that seemed surprising, given its previous statement on prenatal screening, gave a tentative go-ahead to the process of genetically modifying human embryos prior to implantation. The report said that the process would be morally permissible if it was in the child's best interests and did not contribute to already existing social inequalities. The obvious use would be to edit genes that would otherwise lead to serious diseases or conditions, but the terminology used to define 'best interest' is wide open to interpretation.

In recent years there have been incredible advances in gene-editing technologies. The CRISPR/Cas technique, which was inspired by bacteria's mechanism to protect their genome from viral attack, can be used to alter genes in any organism. There is widespread support among the scientific community for using CRISPR/Cas in the context of research but some scientists, including Professor Jennifer Doudna, one of CRISPR's co-inventors, have urged a worldwide moratorium on applying it to the human genome for clinical use, as in the case of treating embryos prior to implantation, until the full implications have been 'discussed among scientific and governmental organisations'.

Aside from the ethical and moral questions, it is still not known whether the technique will prove safe for individuals long-term.

As this book was going to press a rogue researcher in China claimed to have used the technique on embryos to genetically edit them to be HIV-resistant, and then implanting those who were successfully edited into their mother's womb. The twin girls were supposedly born in 2018. The worldwide scientific community was left reeling at the audacity of ploughing ahead with such experimental techniques in the creation of human life, without the necessary ethical, safety and legal frameworks in place. The hosting university issued a statement saying it had launched an investigation into the research, which it said may 'seriously violate academic ethics and academic norms.' This case exemplifies the chasm between the speed of technological breakthroughs, enabling experimentation at an unprecedented pace, and careful consideration of how these new technologies should be applied.

As more is discovered about the genetics of behaviour, these matters will become ever more complex and urgent. Will there be a role for society to intervene to mitigate the effects of innate factors in individual cases? We may decide collectively that we would not let anyone suffer with untreated schizophrenia if we were eventually able to identify its full range of causes, but would we say the same about low intelligence? How about autism, ADHD, or mania? Some individuals diagnosed with these conditions point out that they come with positive aspects – increased resilience during times of social turbulence, the ability to view the world in a logical and systematic way no matter what is thrown at them, creativity and enthusiasm for adventures, periods of high productivity and relish for life. Should these associated traits also be wiped out? All these conditions are linked with multiple genes so, for now, it would be impractical to edit them out of existence, but that will change as the technology becomes more sophisticated. As the Nuffield Council emphasises, public discussion about how these technologies should be applied in the future is essential.

When we look beyond health to social outcomes the picture becomes subject to political as well as moral considerations. During the course of researching this chapter I spoke to Dr David Hill, a statistical geneticist at Edinburgh University, whose work suggests that the genes implicated in higher intelligence also correlate with longevity, greater happiness and higher life-time socioeconomic status (SES). If there is a small but significant genetic component to these important aspects of our lives, as his work suggests, should that change our collective conversation about intervening to reduce the kind of poverty that is handed from one generation to another? Studies have already indicated that growing up in a low socioeconomic environment correlates with disadvantages to neural development. Unfortunately, it's easy to see how this new genetic knowledge could simply be used to reinforce social inequalities rather than try to create systems that help mitigate them. The danger is that politicians and others use biology as an argument for non-intervention.

As always, though, data can be used in many different ways. David believes that his results, and others from the field, could be used to support intervention to reduce social inequality since they supply a useful way to measure the scale of the problem and the effectiveness of any intervention. 'The level of heritability for something like SES is an indicator of how equally the environmental opportunities for success are distributed across a society,' he told me. 'So a higher heritability for SES would indicate a more equal environment.' In other words, our ability to identify genes linked to higher SES might actually signify the fact that we are creating a more equal society. We'll be returning to the details of this research and its potential ramifications later.

Swerving the fate of suffering

We are now advancing into an era of medicine in which an individual can receive treatment tailored to their unique profile. A combination of technological innovations, including whole genome sequencing, wearable tracking technologies and rapid high-throughput analysis of big data sets, is bringing about an era of predictive and personalised care. We will soon be able to predict how each patient will react to specific interventions as well as identify risk for developing specific illness. Clinicians, pharmaceutical companies and policy-makers are steering away from the old one-size-fits-all approach to both patients and their conditions. There is less expectation of finding the magic-bullet/ wonder-drug that will treat a disease in everyone, in favour of embracing this fact: although as a species we share many commonalities, as individuals we are unique, in sickness and in health. The expectation is that in the future it will be possible robustly to predict the health fate that awaits us and use precise methods to identify the treatment best suited for our individual body chemistry before we fall ill. The possibility of thwarting biological fate before it has manifested itself is tantalisingly close.

This is precisely the scenario I wished for almost two decades ago, when I was working at the psychiatric hospital as a nursing assistant. It was painfully obvious even then that psychiatry was in dire need of reinvention. The diagnostic system had obvious failings in terms of specificity, sensitivity and outcome. Often it was not clear what a patient was actually suffering from. Psychiatric diagnosis lacks the sort of sensitive tests that other branches of medicine routinely deploy. If you suspect somebody has an abnormal thyroid gland, for example, you measure thyroxine levels and adjust them accordingly, through prescription of hormones or surgery. Psychiatric diagnosis, in contrast, is based largely on how the patient reports they are feeling.

There are obvious problems with this. For a start, experience of wellness and illness changes from day to day. In addition, some people with severe mental-health problems may not have the cognitive ability to analyse their emotions or even identify what is real, let alone communicate their inner state. As a result, if you ask two psychiatrists to diagnose a patient they will agree with each other only approximately 65 per cent of the time.

To complicate things even further, a diagnosis, whether of schizophrenia, autism, bipolar disorder or depression, did little to predict how the people I worked with would progress in life. There was so much overlap in symptoms between conditions, and so much variety in presentations and life trajectories among individuals, that diagnosis frequently provided very little reliable information upon which to base treatment or work out a prognosis.

Even once a diagnosis was in place, treatments were often neither targeted nor effective. They were reliant on a handful of drugs that had many serious side effects. Pharmacology had largely replaced lobotomy and electro-convulsive therapy back in the 1960s, and for a few years there was jubilation in the medical world as doctors believed science had presented them with a range of miracle treatments. Unfortunately, though, the intervening years have demonstrated that pharmacology is not without its own problems.

The development of drug treatments coincided with and was driven by the groundbreaking discovery of the chemical synapse. Psychopharmacology involves a patient taking a drug to either activate or block receptors for particular neurotransmitters at the synapse, thereby controlling the passage of information across the brain. Unfortunately it is impossible to use drug therapies to target just one brain system. Neurotransmitters do not operate as a key to open a single pathway in the brain. They are much more multi-tasking than that and are able to 'talk' to many different nerve cells by fitting into a variety of different receptors. They have different effects, depending on which type of receptor they lock into. Adding to this intricacy, receptors are widely expressed throughout the nervous system. Essentially, there's no way that swallowing a tablet will lead cleanly to a change in a particular behaviour because the active ingredients have a widespread action across the entire nervous system. Which is why all drugs come with side effects.

Pharmacology has made advances since the sixties and drugs have been developed that help some patients with some symptoms, but the idea of 'curing' psychiatric disorders with new drug treatments has stalled. In the meantime, patients are still suffering with debilitating side effects, which are highly personal depending on how an individual expresses their receptors, their existing chemical make-up and their metabolism. As a result, most people with mental-health problems have to go through an exhausting process of trialling different drugs of varying doses before they find one that may help them, and even then they must tweak their treatment on an ongoing basis: their receptor expression profile and sensitivity to the chemical may change over time.

During my time working at the hospital things were pretty bleak but there have since been significant advances in the understanding of causality of mental-health problems, with hopes raised for improved diagnosis and treatment. There is a growing body of evidence to suggest that the majority of psychiatric conditions are neuro-developmental in origin: the problem starts with a fault in the way neurons are wired together as the baby develops in the womb. Genetic factors play a role in this, and prenatal environment is also significant. Prolonged exposure to high levels of maternal stress hormones, maternal infection or severe substance abuse may contribute. Once the child is born, other experiential factors during early infancy may also be significant, as we saw earlier.

To find out more about the genetic factors contributing to psychiatric and developmental disorders I spoke with Dr Kate Baker, clinical geneticist at Cambridge University and Addenbrooke's Hospital. Kate's specific focus is on finding the genetic cause of neurodevelopmental problems in children. The families she meets are referred to her by their GPs, often after months or years of worry about their children. With new genetic-testing methods, such as chromosomal microarray and exome sequencing, it is increasingly possible to identify genetic variations associated with a broad range of diagnoses, including autism-spectrum conditions, schizophrenia and learning disabilities. However, interpreting these test results for individual patients and families can be complex. As a clinician, Kate spends a great deal of her time discussing their meaning and creating tailored care plans.

I described the problems I'd encountered when I worked at the psychiatric hospital and asked if things had changed since then. It was wonderful to be told that Kate had a much more positive view. It seems there is now a definite and growing acceptance that people don't fit into a specific box that labels them in a particular way for ever. 'We are all a complex mix of body, brain and social experience,' said Kate. 'I work with patients who might fit into many different boxes – say, a child with autism, epilepsy, dyspraxia and learning disabilities. My job is to put all that to one side and ask whether we can understand the mechanisms that link or cause the variety of symptoms that come and go.'

It is now known that a proportion of patients in a psychiatry unit, though not all, will have an identifiable genetic problem that is a significant contributing factor. For example, chromosomal tests can identify small but important differences (deletions or duplications of genetic information) in around 10–20 per cent of autism and learning-disability cases and 5–10 per cent of schizophrenia cases. Health services have been slow in adopting these tests, though, largely because they don't yet impact on treatment. The information can't yet be used to predict exactly how conditions will develop or how the patient will respond to particular therapies. In theory it could be used to shorten the sometimes distressing process of trial and error of different treatments, but for most genetic diagnoses the evidence to back up personalised treatment options is not yet available. One significant benefit of diagnosing a genetic disorder is that it enables doctors to identify risk factors for physical-health problems in later life. In practice, though, receiving information about adult-onset medical problems can be yet another source of worry for families with young children, whose concerns in the here and now are about learning and behaviour.

If the impact on treatment is still limited, testing can make a positive difference to how families care for their children. Some of the children Kate works with carry a genetic predisposition that means they struggle to process and filter information. Kate and I discussed heartbreaking stories of sleep-deprived parents struggling with a toddler who becomes severely aggressive and self-harming when they go to the library. For some children the fluorescent lights, the echoes and reverberations, the bright colours all act as sensory overload. The parent can find themselves utterly bewildered, desperate but unable to help. 'It's a huge relief to these families to be told there is a physical factor,' Kate told me, 'and to hear that they can support their child by avoiding certain environments and providing extra reassurance and time for adjustment when their child encounters transitions in life.'

Currently the most significant benefit of knowing about a genetic factor, interestingly, seems to be psychological. Kate told me that for many of the people she deals with, knowing there is a genetic alteration underlying the behaviours that have been so troubling can be extremely reassuring. It helps to address some of the stigma associated with psychiatry by allowing people to comprehend that they are dealing with a disease of a biological component. The test results can ease the families' sometimes overwhelming sense of personal responsibility and help them move towards greater acceptance. Once everyone in the child's orbit has accepted that the child can't simply 'snap out of it', they can start to accommodate the condition in the same way they might, say, if their child had juvenile diabetes. On the other hand, there is often disappointment about the lack of information available and frustration that finding the genetic cause does not change treatment. There are still many unknowns about each individual's future.

What about the more nebulous matter of using genetic diagnostic tests to predict the appearance of a disorder in someone who has not yet exhibited symptoms, or even been born? I asked Kate about a scenario in which a family might use this technology to inform their decision about having a second child when the first has a severe genetic psychiatric condition for which there is currently no cure or efficacious treatment.

Before she would entertain discussing this scenario, Kate was keen to point out that the genetic tests currently predict susceptibility to particular conditions: they are not reliably predictive of outcomes. Every person's case is different. That said, she works with parents who are in precisely this position, and the process of supporting them to explore genetic testing on an unborn child is extremely delicate. 'The first thing to bear in mind would be the severity of the first child's problems. That would inform the whole conversation. We might offer to test the parents to see whether their first child had a de novo mutation [a completely random and unfortunate switch in the child's DNA during early development], or whether it had been inherited from a parent. If one of the parents was a carrier there would be a 50 per cent risk of any baby inheriting the gene mutation. If we're talking about a high-specificity test for a severe outcome, then the family might want to consider testing the unborn baby, and if they did, that testing would be supported. If the results came back positive we would discuss all options with the parents, including termination. If a family decide to continue with a pregnancy despite a positive result, the knowledge can smooth the way to earlier treatment, as well as help them to manage expectations and tailor their environment to the child.'

Kate works with people who find themselves in desperately hard situations, and though she is committed to her work and knows the good it can do, she also stressed that genetic testing in this context poses difficult questions. 'I'm the first to say we should be doing more genetic testing but it's a very sensitive matter. You have to work hard to prepare people, ask them a lot of "How are you going to feel *if* such and such . . ." questions. I've worked with people who think they'll be able to handle diagnosis of an unwanted outcome in a calm way and actually get more anxious once they have it. It can affect their engagement with that child. And, for example, we don't advise testing babies at birth for these conditions because there are real questions over whether it's helpful to an individual to carry a label that says, "I'm susceptible to X or Y condition." That can do more harm than good to a child, leading to them feeling very alienated from their family and peers. That's why we wait until a problem emerges.'

My conversation with Kate flagged up the ethical complexity that comes along with our greater understanding of how our neurobiology shapes our fate. It set me on a path to engage more deeply with the tougher implications of the increasing predictability of our brains. There are relatively few instances where biology truly is destiny, but when it is, that destiny is a dark one indeed.

Up against the limits of knowing

Huntington's disease offers an unusually dramatic and clear-cut template for assessing the impact of knowing one's fate. This disorder arises when a mutation in a person's genetic code dramatically affects how their neural network is built. It has been widely studied, partly because it arises from a single gene change so it lends itself to investigation, but also because it is such a severely debilitating condition for which there is no cure. If you have the mutation you will definitely develop the disorder. Fifty per cent of children of those affected inherit the mutation and go on to develop the condition, so for afflicted families, it can be devastating. There is a wide spectrum of symptoms, including uncontrolled movements, poor coordination, a progressive depressed mood, anxiety, irritability, apathy and psychosis.

People are typically in their thirties when they begin to show symptoms, and die fifteen to twenty years after diagnosis. The root cause is a single gene, called HTT, or Huntingtin. The protein it codes for is involved in brain energy and connectivity dynamics. A huge amount is now understood about the precise cause for Huntington's but unfortunately there is currently no cure, although progress is being made on a weekly basis and treatments are improving. Genetic screening for the condition is available on the NHS for the children of affected people. But opting to have the test comes with significant considerations.

To find out more I spoke with Lizzie, whose father was diagnosed with Huntington's in his late forties. For a long time he was able to live a relatively symptom-free life but his condition gradually worsened, and two years ago, when he was in his early sixties, he 'started waking up in the night with psychotic episodes. One night my mum called the psychiatric team, who ended up sectioning him. He has been living in a secure nursing home ever since.'

Lizzie's experience of Huntington's is unusual since her father developed the disease relatively late in his life and his decline has been slow. Her childhood was unaffected by his condition and she had already moved out of the family home by the time her father started showing symptoms. But over the following twenty years she has experienced, first hand, her family responding to the condition, with the knowledge that she also stands a significant chance of developing it. After careful consideration, she decided not to be tested. She has a 50 per cent risk of carrying the faulty gene and, if she tests positive, stands a 50 per cent chance of passing it down to her children. Has this potential risk affected her decisions about raising a family?

'My partner and I did decide to have a family,' she told me, 'but we chose to have our children young. Given that my father has late-onset Huntington's, we decided to assume that if I were to get ill, our children would be more or less grown-up by then, and to hope that treatment will have advanced considerably. I felt that by the time my children are of an age when they might develop symptoms, the advances in treatment will be amazing. It's been hard, but the whole experience has made me focus on making the most of life. When Dad went into the nursing home I turned down a job I would have loved to take and we moved back to Manchester, so I can visit him more often.'

I also spoke with Maria, who is forty-three. Members of her family started to be diagnosed when she was in her mid-teens, with her mother exhibiting the early-onset and severe variety of the disease. 'My mother's diagnosis was horrible. I was terrified for my own future, but the massive thing for me was deciding that I wouldn't have children because of the risk of passing on the illness.'

Maria postponed taking the test for many years. She wanted to have it, she received counselling for the process, booked the appointments and travelled to the test centre a number of times, but having watched her mother's illness progress, she struggled to go through with it. Each time she backed out of the process before receiving the result. 'I couldn't imagine how I would live my life, knowing I was going to get the disease. Some people I talk to say flippantly, "Oh, I would just have the test. I would have to know." But until you're in the situation, you have no idea how you're going to feel. Or how terrifying it is. On a basic level it would have made getting a mortgage or life insurance, the normal things you take for granted, much more complicated.'

Eventually, two years ago, Maria went through with the test. She had not developed any symptoms. She told me that she 'could finally face doing it because I was thinking I probably didn't have it'. Her test, thankfully, came back negative. 'The relief was amazing.' Unfortunately, though, by the time she got the news she was unable to have the children she had desperately wished for. That was why she was speaking to me, advocating more discussion, openness and less stigma around psychiatric disorders, so that people know they are not alone. Even with rare diseases, such as Huntington's, there is a community of people with similar heartbreaking stories who would like to reach out, as well as organisations, like the Huntington's Disease Association, which put me in touch with both Lizzie and Maria, and offers support as well as information on new research.

I came away from my conversations with Lizzie and Maria

profoundly moved by their stories and with a more nuanced take on my previous stance that it is always better to know one's fate. My experience of being told I was a carrier for haemochromatosis, and the ambiguous position in which that news placed my son, had temporarily rocked my previous faith in the adage that knowledge is power. It was now clearer to me than ever that, as with almost everything in life, that depends.

Lizzie has opted not to take the test and has still not reached an age when she could confidently say that she is not going to develop the condition, but she has suffered a lot less from not knowing than Maria. Lizzie's particular family circumstances, with late onset and slow advance of the disease, as well as a very supportive partner who shouldered the risk with her, enabled her to take a decision that wasn't open to Maria, whose mother had the severe variant. For Maria, not knowing was agony but it was still preferable to certainty, if that certainty was the outcome she dreaded. Talking to them was a sad reminder that, however much progress has been made in understanding the mechanisms of causality, until there is a cure Huntington's is a grim fate indeed.

Mercifully, the question of our biological fate is very rarely as clear-cut or as severe as it is with Huntington's. Most health outcomes are multifactorial, allowing some wiggle room for an individual to react to the news and take action that might improve their outcomes. A recent breakthrough in the early detection of plaque build-up linked to Alzheimer's raises just this possibility. In 2018 scientists based at the Shimadzu Corporation in Japan and the University of Melbourne, Australia, revealed that they had developed a simple blood test that accurately predicted in 90 per cent of cases the build-up of particular protein deposits, or plaques, in the brain. These plaques are strongly linked to the development of Alzheimer's, one of the scourges of our ageing population, affecting up to 50 million people worldwide.

The exciting thing about this test is that it can predict who is likely to build up dangerous levels of plaques up to thirty years before the first symptoms of Alzheimer's emerge, allowing people to take steps to change their lifestyle in a way that may reduce the impact of the disease. The scientists involved in the work stressed that we still don't know exactly what causes Alzheimer's, and there is as yet no cure, but an early-warning test like this could help on both counts. Early detection could be of benefit to the individual if it allows them to take steps to mitigate the disease's progress, to access treatment earlier, prepare themselves and plan for care. The test could also be used to screen people to participate in new drug trials, with the hope of speeding up the progress towards more effective treatments and even a cure.

While it's impossible to say for sure what causes the plaques in the first place, we've already seen that a number of correlation studies show that regular physical exercise, eating a varied diet and getting at least seven hours' uninterrupted sleep a night are all linked to maintaining healthy brain functioning into later life. So, while it wouldn't be quite right to say that this blood test will offer a way to swerve our fate, it's not too much of a stretch to say that it puts us on that path. It certainly looks like a big step on the way to simple, affordable personalised medicine that empowers people to take the steps that will most benefit them before problems emerge.

Why can some people dodge their fate and others not?

Although we are learning more and more about how environment and biology interact and are implicated in some psychiatric conditions, we still find it hard to predict who will develop a condition and who won't. Why does one sibling develop chronic depression in response to childhood trauma, say, and the other, seemingly miraculously, emerge unscathed? This is the question that got me started on neuroscience all those years ago, and I've found myself looping back to it many times since.

There have been a number of interesting neurobiological studies recently on the subject of resilience, which for this purpose is defined as the ability to maintain a healthy outlook despite experiencing adversity. Children who have experienced emotional, physical or sexual abuse, for example, about which a lot is known because, unfortunately, it is relatively common, are significantly more likely than children who haven't had this experience to go on to develop mental-health problems, including addiction, self-harm, antisocial behaviours, depression and anxiety. But not all of them do. Between 10 and 25 per cent grow up to lead normal, healthy lives. So what's different about them? Did they have an adult who cared about them, even if that person couldn't prevent the abuse? Did they have a friend, or faith, or higher self-esteem?

It transpires that resilience is complex, but it does have a genetic component. One of the genes implicated is brain-derived neurotrophic factor (BDNF). It produces an extremely useful chemical that helps to support the survival of existing neurons, encourages the growth of new ones and cultivates connections between them. A variation of this gene, Val66Met, instructs for very high levels of BDNF to be expressed. Individuals with this genetic variant have a flourishing brain. Their hippocampus, a brain region involved in learning and memory and one of the few areas in which new nerve cells can be born throughout life, is larger than in most other people's. This is linked to the ability to create and store new memories easily, forge new frameworks for thinking and be flexible in how you view and experience your life. It makes sense that an individual carrying the gene that instructs high production of BDNF would be more resilient than one without, and in fact the 10–25 per cent of children who have experienced abuse or neglect but don't develop mental-health issues are much more likely to have this BDNF coding.

But even if we restrict ourselves for a moment to the genetic contributory factors, it's not as simple as identifying a particular BDNF variant as the single gene for resilience. For a start, as we have seen throughout the book, our genetic code is by no means the sole agent in causing behaviour. Genes respond differently in different environments, their volume dials turned up or down by environmental triggers. And as we know, with a complex trait such as resilience, numerous genes are likely to be implicated in its occurrence. Some of the others (doubtless more will be discovered) include having a long version of the gene that instructs for the serotonin transporter (5-HTTLPR(SLC6A4)), which in addition to its vital part in the effectiveness of serotonin, which is involved in generating feelings of happiness, is also linked to reduced amygdala reactivity. A third gene contributing to an individual's coping capacity is Neuropeptide Y (NPY). Unfortunately, if you have the genetic variation called rs16147, your amygdala is more likely to be hyper-reactive and you are therefore more likely to feel fearful and anxious.

Genes involved in regulating the inflammatory response to stress also play a part. If you have a specific variation of a gene called FKBP5, for example, it lessens your chances of attempting suicide or developing post-traumatic stress disorder. And that's just genetics. Then there are the other contributory factors.

To discuss resilience further, I spoke with Dr Anne-Laura van Harmelen, another scientist based at the University of Cambridge, who has been investigating how and why some children and teenagers respond with resilience after adversity.

Resilience, when it's understood as a biological concept, is highly complex and encompasses many different behaviours in response to tribulation. There are some overarching themes, though. If, for example, you are unfortunate enough to have an unlucky repertoire of genes that predispose you to social anxiety, impulsivity and poor emotional regulation, and you also experience a significant stressor, such as abuse, injury, illness or abandonment, you may be more likely to hurtle down a path triggering a cascade of powerful environmental and social factors that will further damage your mental health, which in turn perpetuates any genetic predisposition. Anne-Laura is interested in trying to discover more about how we can protect resilience levels at a group level and, critically, boost resilience for the individual by understanding our biology a little better: then we can target our inherent weaknesses in a more personalised way.

'Resilience is not a static thing you have or don't have,' she told me. 'It's dynamic. It's about resilient *functioning* and is facilitated by many interrelated factors. So it may be that your genetic predisposition means you have better emotional regulation skills, which in turn might mean you respond better to interpersonal events, which might make you nicer to be around, which means people want to hang out with you, which reduces your stress levels. All of these things are important and influence one another.'

Anne-Laura was keen to emphasise that we should be cautious about leaping to conclusions from the field of resilience to date. 'Many of the neurobiological studies into resilience have been carried out on small samples. Meta-analysis demonstrates that we need more data before we can make more robust conclusions.' The genetics of mental-health resilience is a field that is particularly vulnerable to hyperbole, partly because its mechanisms are highly complex, and over-simplification tends to lead to incorrect assertions about causality. 'It's best to think of genes as leaves on a tree,' she said. 'They're important but the influence of any single leaf on the shadow cast by the entire tree is small.'

It was fascinating to talk to Anne-Laura about the broader context for resilience that she had observed in the lives and personalities of the people she works with. Some of the strategies that resilient people use to cope are potentially useful to all of us, irrespective of our genetic heritage or early-years' experience. They include an ability to distract yourself, to take control of your thoughts so that you don't get upset when provoked, and to avoid dwelling in negative ruminations. Being able to recognise and name positive events, however small, is also crucial, and can be cultivated. Studies, admittedly in rodents rather than humans, have shown that exercise, the opportunity to explore new environments and to socialise increase levels of BDNF, as does proper sleep. No surprise, really, given that these are all behaviours that have been conclusively linked to better wellbeing in people across a number of different areas of study.

Higher self-esteem is a resilience factor, as is a supportive and positive family and friendship group. Anne-Laura's work has, interestingly, shown that family support at fourteen years old predicts friendship support at seventeen. A child acquires from their family a template for how to interact with and support people, which seems to dictate their expectations of, and interactions with, their friendship groups. It's not derived solely from the interaction between caregiver and child: it's also shaped by the child's observations of how the caregiver interacts with their own friends. It takes a village to raise a child, as they say, and the child learns its skills, including how to be a friend, from the village.

Predicting behaviour rather than health

Even for most health conditions it is still difficult to say for sure who will succumb and who will not. For complex behaviours like resilience, the neuroscience is in its infancy. That is changing, though. As further research is conducted, no doubt more will be discovered about how people are born predisposed to specific behaviours and traits, and as that research feeds into other disciplines, such as psychology, sociology and economics, we will understand more about the extent to which our brain generates our decision-making, our personality and even particular events.

The first data sets from huge genome-mapping projects are now becoming available and there is a rush to mine them for useful insights. Even the most rigorous of this work is provisional, but to get a feel for how we might soon be able to predict individual outcomes in a context other than that of health, I spoke to Dr David Hill of Edinburgh University, who has conducted a meta-analysis of all the studies into the links between genetic factors and intelligence.

The first thing I wanted to know was what intelligence means in this context. David reassured me that, though the popular perception of IQ tests, for example, is that they measure competence only in whatever the test is targeting, intelligence has now been so widely studied over such a long time that its meaningful measurement is totally possible. This is achieved by combining results from a number of tests, of memory, verbal and numerical reasoning, and reaction time. It's been demonstrated that the results from these tests correlate, so if a person scores well in one they will score well in the others. 'So, if you wanted to predict how well someone will do in school, for example,' David told me, 'you'd run a battery of tests and look for the amount of correlation, the common factor. That commonality is what we define as intelligence.'

The focus of David's work has been to try to disentangle genetic from environmental factors to say something about the relative significance of genetic factors that act on intelligence. His conclusion was that 'Genetic effects account for half of difference in intelligence between individuals and around 40 per cent of educational differences.' In a separate study, David found more than five hundred genes linked to intelligence differences, as well as the first evidence that the genes involved in the biological process of neurogenesis, by which new brain cells are created, seem to be of particular relevance for intelligence.

The identification of multiple genes involved in a complex behaviour has been seen throughout the book, with Giles Yeo discussing his research into the genetics of obesity, and with Kate Baker using genetics to diagnose psychiatric conditions. We know such traits are highly polygenic, meaning that many genes are implicated in their occurrence. There is a high degree of heritability (the genetic influence passed from parent to child) but it's impossible to say it derives from single genes. As David put it to me, his analogy a slightly wetter, northern ßversion of Anne-Laura's tree, 'If you went out in the rain and came back in and someone asked you which drop of rain made you wet, you'd be justified in saying, "That's not the right question." A more useful question would be "How many raindrops did you get hit by?"'

So, given that David's analysis suggests only 50 per cent of intelligence derives from genes, where's the rest coming from? 'There's a huge amount of environmental influence. Education will account for some of it. That might sound intuitively obvious but we now know that education helps boost intelligence and vice versa: that intelligent people stay in education longer. When the UK school-leaving age was raised by a year in 1972, intelligence levels among the general population subsequently went up.'

As we saw earlier, education also serves a protective function for the brain, helping to keep it healthy for longer. So that's one thing under our control that we can do to help ourselves. Or is it? David's research also indicates an overlap between the genes involved in intelligence, longevity, income and susceptibility to psychiatric conditions, and these genes are also linked to likely participation in higher education. So, is it that the people with a certain genetic constitution are also those who are more likely to stay in education and reap the benefits it offers the brain? Is it all a self- perpetuating, self-reinforcing loop? Perhaps, if our choice to stay on at college or not is also a result of our inherited traits, then altering our genetic destiny may depend on external intervention. This certainly gives context to the tendency of states all over the world to raise the school-leaving age and might boost arguments in favour of a 'nanny state' where policies help nudge us in a more favourable direction of behaviour whether we personally desire it or not. Is it possible to imagine behavioural traits following patterns of physical characteristics and changing across generations? Height, for example, is made up of both genetic and environmental input and has shown a trend for average increases across the majority of the globe over the last few centuries. Genes are linked to height and it is widely accepted to be a highly heritable trait, but the speed at which average height has increased cannot solely be attributed to natural selection favouring taller mates. Rather, the nutritional content and amount of available food has generally improved, which might contribute to increases in height. There is still a distribution, of course: those individuals genetically predisposed to be taller still are, while those destined to be smaller retain that attribute; but, as a whole, the population has shifted slightly but surely towards the upper end of the scale. Perhaps, then, might we likewise be able to nudge certain aspects of how our minds and our behaviours operate?

Meanwhile, there are companies leaping on preliminary findings emphasising the importance of genetics and attempting to commodify them for the current age of anxiety-ridden parenting. US company Genome Prediction, for example, offers embryo selection based on analysis of numerous genes linked to intelligence to those who can afford the price tag of the service.

Robert Plomin, psychologist and geneticist at King's College, London, who is best known for his work on twin studies, is the author of the controversial *Blueprint: How DNA Makes Us Who We Are*. He argues that DNA can be viewed as a 'fortune teller' that can 'predict your future from birth'. Kevin Mitchell, geneticist and Associate Professor at Trinity College, Dublin, also recently published *Innate: How the Wiring of Our Brains Shapes Who We Are*, late in 2018, the central thesis of which runs along similar lines to Plomin's, agreeing that behavioural traits such as intelligence are composed of strongly hereditary elements, and that although there is still a role for environment, both factors could, plausibly, be modified to alter an individual's life trajectory. But Mitchell also discusses a third route by which variance in behavioural traits can be offered, one that is, as yet, intangible to human manipulation. He writes:

The wiring of the brain is astonishingly complex and its almost miraculous self-assembly relies on a huge number of cellular and developmental processes, involving the actions of thousands of genes. It is variation in precisely those kinds of genes that has been implicated in intelligence.

The idea is that since the development of the neural circuitry involves billions of nerve cells, trillions of connections and many more factors of biochemical interactions between protein molecules, the level of potential background noise or inherent randomness at a molecular level that can occur during its construction is vast. Admittedly the genes set the rules for how the system will be created, but fuzziness creeps into the biological system due to its complexity – and this is particularly true for the brain – which means that small differences are then amplified as development proceeds. Kevin argues, 'We can certainly use genetics to look at statistical effects across the population, but this will give at best very fuzzy predictors for individuals.'

But, moving away from scripting the intelligence of future generations to the patients at the psychiatry hospital I worked at two decades ago, I desperately hope that biological markers, including genomic ones, continue to be identified and that they may help to pinpoint or predict psychiatric illness early, even with the complexities involved.

Analysing the vast data sets becoming available, and disentangling the various influences on behaviour they illuminate, is highly complex work that will be ongoing for years. Given the current revolution in technologies, more and more data from studies examining these themes will emerge, and as David says, that 'data could be used in any number of ways to support different interventions, or to resist them'. We urgently need to debate these issues because, although they are challenging, they affect millions of people.

A study led by scientists at the University of Pennsylvania and the University of California, San Francisco, demonstrated that by the age of five, a child's brain has been literally shaped by their socioeconomic status. The poorer the child, the more reactive their stress responses, the thinner their frontal cortex, the poorer their frontal function, covering working memory, emotional regulation, impulse control and executive decision-making. This is the sort of scenario we will need to grapple with, as we understand more about how people's outcomes are determined by neurobiological factors beyond their control.

So is the future bleak or bright? When it comes to neurobiology it's both, because it's all there for the shaping. There is a risk that, as we identify more and more biomarkers for everything from congenital conditions to creativity, we will end up with a society of the genetic haves and have-nots. If personal genetic information is commodified we might price people out of the social contract or consign them to second-class-citizen status at birth. Perhaps we are entering a Brave New World. As a society we have to be ready to imagine the future and grapple with the implications.

It seems as though we are swinging on a pendulum between acceptance and prevailing will. In the early 20th century the belief was that many aspects of our character were ingrained, immutable. This view brought with it the atrocities of eugenics, horrifying millions across the globe. By the late 1990s, the pendulum had fully swung its arc and the concept of brain plasticity was in vogue, both in the scientific world and the general zeitgeist, when the world seemed to be opening up to the limitless possibilities in communication, technological and personal development, the idea of limitless potential. But the pendulum swings again and we now appear to be reaching full oscillation: there has been a resurgence in the belief that perhaps our brains are not as plastic as we had hoped, and that although our brains may not be hardwired *per se*, they might be pre-wired for a certain trajectory in life, a way of viewing the world and how we respond to it. It is this idea that we will increasingly have to wrestle with as vast datasets from the study of the connectome, genomics and proteomics avalanche towards us in the coming years.

Developments in neuroscience will certainly have beneficial effects on the lives of millions, though. They will be felt first in medicine but will be incorporated into other domains as well. All those people struggling with psychiatric disorders and with Alzheimer's might expect to see massive improvement in their quality of care and their quality of life. As breakthroughs in other disciplines intersect with those of neuroscience even more progress will be seen. Recent advances in understanding the interactions between gut and brain, for example, mean that analysing and tweaking gut functioning can help predict, diagnose and treat psychiatric conditions, those diseases traditionally thought to be simply of the mind. The same goes for interaction between immune system and brain.

Vigilance is essential as we grapple with the dilemmas posed by advances in neuroscience, but there are many reasons to feel optimistic. The mother of one of my neighbours, in her late fifties, was recently diagnosed with Parkinson's disease and opted to have groundbreaking brain surgery based on the technique of deep-brain stimulation that we encountered earlier. Her skull was opened and an electric-shock panel was embedded in a specific circuit of her brain. She receives zaps of electricity at a particular frequency of oscillations, which effectively switch off her previously debilitating symptoms of tremors and depression. Her operation took place six months ago, and it has been a complete success. She now cycles cheerfully around Cambridge and has gone back to being involved in local activism. Parkinson's disease can run in families, but her son has been tested and fortunately seems to be clear. His mother's life has been reclaimed and he does not have to live in fear of developing the condition himself. This is great progress and there is much more to come.

CHAPTER EIGHT

The Cooperative Brain

The previous 150 years have brought staggering medical advances for humanity: antibiotics, adoption of aseptic surgery and reliable contraception to name but a few. Although this has meant we are no longer at the mercy of our biological fate to anything like the earlier degree, the chinks in our organic armour remain a constraint. The latest neuroscience breathes new life into the concept of fate by situating it right at the core of what we increasingly believe makes us who we are: our brain. It shows us that we are born with our behavioural propensities already sketched out. Neurobiology and environment then work in tandem to keep us inking in the lines of the sketch, finessing the detail to produce the full picture of our life.

In seven chapters we've been following the thread of an argument that runs something like this: the brain we are born with, the product of millennia of evolutionary forces, is subject to many species-wide characteristics but shaped by our unique genetic blueprint. Our early years tend to cement our proclivities through considerable exposure to an environment that has typically been created by our parents, from whom we inherited those tendencies in the first place. Given that our brains filter information on the basis of prior experience, our present and future reality is based on an avalanching amplification of what has happened before. We end up stuck in the version of reality that our brain has constructed, which is a cramped simulacrum of the vast external world, a simulation that has moulded itself to fit our expectations. Much of the time we dwell there quite happily but occasionally we feel its constraints or notice its glitches. We are unhappy and can't shake the mood. We try to change a habit and may struggle for a while before giving up. Or we try to change somebody else's mind, which usually doesn't end well. We bump up against the constraints that govern our selfhood, other people's and our version of the world. The experience can be frustrating. Misunderstanding, resentment and aggression may result. That's just human nature, you may say. We're fated to struggle with ourselves and each other, and most of the time we struggle in vain.

The myth of innate human nature

Of all the things that our fallible brains love to generalise about, human nature is one of the principal targets. We've come up with any number of grand theories over the centuries and justified them with an appeal to their self-evident rightness. 'You may not like it,' we say to one another (where 'it' could be anything from adultery to humanity's fondness for mind-altering substances and the persistence of social inequality), 'but it's just human nature.' Human beings are the pitiful creatures of the gods, we told ourselves back in the days of Aristotle, or weak and sinful, the prey of Lucifer. Or innately selfish, motivated above all else by passing on our genes, or acquiring more money and status. We evolved to divide our fellow humans into 'them' and 'us', viewing each other as members of distinct tribes based on national borders, skin colour, gender, pastimes and preferences.

Naturally enough there have always been opposing views. One contemporary version asserts that we have finally done away with the notion of fate altogether and have diminished predetermined biology to a minor role. This 'you can achieve anything you want' attitude exclaims that any change is possible if we only put our minds to it, work hard and want it badly enough. Neuroscience has sometimes been used to bolster this argument via the concepts of neurogenesis and plasticity.

My experience of working in a psychiatric hospital inspired me to spend more than a decade using neuroscience to try to get to the nuts and bolts of what gives rise to our behaviour. I knew I didn't subscribe entirely to the view that our biology dictated our life's course. I also knew that I didn't endorse the view, however compelling it may be, that we can all grow into whoever we wish to be. Actually, each of us possesses a balance of genuine constraints and natural skills, and that individuality is to be cherished.

During the course of researching and writing this book I've ended up convinced of something I hadn't entirely anticipated: that there's no such thing as human nature. Not really. Yes, we share species-wide characteristics; sure, biology is deterministic at an individual level to a huge degree, but to say we are collectively this, that or the other is yet another over-simple model. It denies the glorious complexity and flexibility on offer to us as our brains, those billions of unique models of reality, encounter one another. Our brains provide us with a fantastic paradox. We are primed to seek out patterns in our environment; we comprehend the world by categorising, simplifying and making assumptions, based on what we have experienced before. This skill helps us to take shortcuts in information-processing, meaning we retain the phenomenal speed at which our brains can make calculations, judge situations and direct our decisions. Yet it is precisely this architecture of the brain, its flexibility, its intricacy, its dynamism, that means that every person on this planet is a true individual, with the scope to exhibit such wide complexity of behaviours arising from the 100 trillion connections in their mind. This ever-changing landscape of complex neural circuitry, which is driven to seek patterns, ironically means that attempting to categorise people by simplifying their behaviours to binary brackets is nonsensical, since the scope for variation in perceiving reality and complexity of thought and behaviour is so vast.

That said, individual constraints are undoubtedly real. We've just spent the last seven chapters examining how and why we're much more constrained by neurobiology than many of us realise or want to acknowledge. That doesn't mean change is impossible, though, and somewhat counter-intuitively, it may be easier to bring about change at a group level than it is for an individual. We've already examined the idea that at population level we're hard-wired to be social, curious and find exchange of ideas rewarding. The creation and maintenance of a collective consciousness, a vast pool of ideas sustained by a network built on collaborations and relationships, has even been suggested as the ultimate motivating force for humanity. We'll consider the suggestion that altruism is one of many innate human characteristics and that it is not mere idealism to believe that we could collectively foster it to boost our chances of making the changes our world needs.

I am not mounting a counter-argument that says we're all intrinsically altruistic. My exploration of the science of fate has clarified my conviction that human nature is too vast and varied for easy pronouncements of that sort. But just as an individual is predisposed by her genetic make-up to be an anxious person, say, whose experiences in early life may or may not reinforce that disposition, so, collectively, we can choose to shape our world around any number of different beliefs, including the belief that making an effort to open our minds to each other is worthwhile. There is such a thing as the neuroscience of compassion and of cooperation.

Applying neuroscience in the real world

In the previous chapter we saw that the era of personalised medicine has the potential to benefit millions of people, but it's going to require us all to engage with challenging questions about how to manage it for the collective good.

It's this collective good that provides the focus for this chapter. How can new insights into human behaviour be applied to other fields, such as public health, civic life, education and the law? Individually each of us is constrained by our own neurobiology, which plays out in endless ways in our behaviours. We eat more or less junk food, are more or less likely to vote in local elections, more or less likely to lash out in reaction to a perceived insult. Altering any habitual behaviour is, as we've seen, not easy at an individual level. But if macro-level changes are made by altering the environment, through legislation, intervention or planning, we can be collectively nudged towards certain behaviours and supported to maintain them so that significant change happens at a group level. Some people will always choose the kale salad, participate in local democracy or maintain sufficient emotional detachment to avoid ending up in a police cell. Others are much more likely to mainline doughnuts, sit on the sofa on Election Day or punch the idiot driving the car in the next lane over. Most people's behaviour falls somewhere in the middle and is dependent on context. Their proclivities can be fostered or dampened.

The UK government's Behavioural Insights Team, also known unofficially as the 'Nudge Unit', has been suggesting policy tweaks since 2010. Nudge theory is a branch of economics and behavioural psychology that tries to encourage groups of people to carry out specific actions. The guiding principle is to put in place policies that make it easy for people 'to do the right thing' by reducing 'cognitive friction'. It is based on this insight: even when we know something is good for us or strongly suspect it will make us happy, we are reluctant to make change unless it's obvious how we should start and unless it feels achievable. That's our intrinsically lazy and sceptical subconscious brain for you.

For example, Britain's Victorian housing stock is a significant barrier to the UK hitting its targets for reducing carbon emissions. All those leaky old houses haemorrhage heat. The answer is retrofitting houses to reduce heat loss, but it's a costly, and major, job. The Nudge Unit suggested that one way to make it easier for people to install insulation in their lofts would be to offer subsidies for the cost of hiring labour to clear out all the junk lurking up there. By the time the job was under way, individuals would be on the path to reducing their own heating bills and nationwide carbon consumption. Win–win.

When they were tasked to foster public engagement with local politics, the unit performed an experiment with one local authority to see if it was possible to increase the number of people voting in local elections. New voters' names were entered into a draw to win £5,000. Voter registration increased by a small but significant 4.2 per cent. (In case you're wondering, there is no publicly accessible information outlining how those people then voted.) The incentive of a cash prize proved to be a pretty straightforward but highly effective way to leverage the brain's reward circuits in the service of increased participation in democracy, at a relatively low cost for the state.

Some of the most interesting and fruitful work has been in the field of public health, specifically around enabling people to make healthier food choices. You'll recall that our species is driven to crave salty and sweet food and to keep on eating. Now that we live in a culture of abundant and highly processed food, we're in the grip of an obesity epidemic. Despite the multimillion-pound diet industry and the impressive results achieved by some individuals, many more of us find it impossible to modify our behaviours. A recent study, based on observations of 1,500 child and adult members of military families who moved frequently around the world, fuelled the argument that we are not fully in control of our own choices around food. The study indicated that obesity is 'socially contagious'. If you move to an environment where a lot of the people you interact with are overweight, you are more likely to slide down this route yourself. In fact, your BMI will increase the longer you stay there. Context is crucial for amplifying predispositions to certain behaviours.

So, given the unfeasibility of nudging us all to relocate to Chelsea, Tower Hamlets or Brighton (local authorities with some of the lowest incidence of obesity in the UK), is there anything government can do to help us tackle our ever-expanding waistlines? Remember Giles Yeo, the appetite academic, and the strategy he and his wife adopted? They banned their personal weaknesses, pork scratchings and chocolate respectively, from their home. It's standard advice to ensure you don't have any biscuits, wine, crisps (substitute your particular vice here) in the house before you attempt to modify your eating habits, but that requires us to be mindful shoppers and resist the food manufacturers' and retailers' tricks for persuading us to buy these tempting treats.

Enter the Nudge Unit, which not only identified a strategy

to help us all but also worked with the retailers to convince them to implement it. As of 2019, so-called 'guilt lanes' in super-markets will be banned. The aisles and shelves closest to checkout points where customers linger in queues waiting to pay have historically been laden with confectionery. Children mounted a sustained pester of their parents, and the easy availability of sweets at the end of the shopping experience promoted impulse-driven purchases. It was a mechanism designed to seduce the hedonic pathway and override our brain's capacity for impulse control. Not any more. Instead, checkouts are confectionery-free and supermarkets are increasingly handing out free fruit for children. Researchers are creating networks of experimental shops and restaurants so that empirical studies can determine how food displays and promotion can be altered further to nudge consumers towards a healthier diet.

Elsewhere, as food manufacturers invest in neuromarketing to gather more and more information about customers' choices and behaviours, creating new products and crafting adverts in the light of what they discover, public-health campaigners have begun to raise the prospect of litigation. It is illegal to target minors with certain forms of advertising, and this might provide a model for demonstrating that some forms of advertising are so coercive that they shouldn't be allowed. Kelly Brownell, professor of psychology and neuroscience and dean of the Sanford School of Public Policy at Duke University, has suggested that as more research is carried out on the links between children's exposure to advertising and their sugar consumption, there may be possibilities for legal action. There is a growing sense of outrage that manufacturers use their billion-pound might to erode our collective willpower and, when challenged, toss out the argument that obesity is a problem of personal responsibility. Nothing to do with them.

For some commentators any resistance to this on the part of the state via legislation smacks of totalitarianism. They point to the existential risk posed by the state intervening to promote behaviours. It may be fine for a liberal democracy to promote eating more fruit and vegetables, they say, but in a different context it would be deeply dangerous for a government to use neuroscience to manipulate its subjects' behaviours.

This is undeniably true but, given that we now have the tools to encourage such change, it seems obdurate to resist using it for such uncontroversial ends as boosting kids' intake of broccoli. As I argued in the previous chapter, the pace of technological advance is going to have us sleepwalking into an ethical quagmire unless we discuss how we want to use the new science. Debate is essential. Not every issue is easy to take a position on, but we must explore the harder ones if we wish to avoid unforeseen consequences. We must open our minds to the idea that the science could be a force for benign manipulation, not least because it is already being used for nefarious ends, as the actions of some food manufacturers or the Facebook scandal over data misuse demonstrate. In this context, the Nudge Unit seems to me a valuable force for encouraging behaviour change. The caveat is that it must operate with total transparency and that it should be tasked with investigating areas that are genuinely of public concern. For it to carry out its work in ways we can rely on, we need to ensure we've had a collective conversation about the kinds of behaviour change we want to see.

There are other potentially fate-changing (and relatively uncontroversial) applications of neuroscientific research to be made in the field of education. Educational neuroscience is growing fast, fuelled by excitement at understanding and harnessing the potential of those periods of massive neural activity during a child's early years and adolescence. Much of its focus is broadly to understand how genetic predispositions manifest in an individual's brain and how they can be nurtured through education and upbringing. Work with twins has demonstrated the role of genes in reading and mathematical ability, where multiple genes converge to frame a person's skill level. Those genetic factors interact with environmental matters, such as diet, exposure to toxins and social interactions, to give rise to the individual's overall ability. These studies provide another crucible for furthering our understanding of how complex traits, such as intellectual ability, arise and are embedded or stifled by potentially controllable factors in the environment.

As more is understood about the biological basis for conditions such as dyslexia and dyscalculia, for example, more effective strategies for supporting individual children are developed. Recent studies have shown that awareness of the rhythm and sound system of language is the best predictor of reading ability in all children, which helps to explain why traditional nursery rhymes, singing, dancing and marching games are such highly effective learning tools. Dyslexic children seem to be relatively insensitive to rhythm in any form of sound. They just don't perceive it in the same way as other children. The suggestion is that teachers increase their use of music-based learning and physical activity to build up the children's awareness of rhythm. Initial findings suggest this can have a powerful impact on reading ability in later life and it has the additional benefit of being play-based and non-judgemental for the child.

Some insights from neuroscience may have a fundamental impact on highly practical aspects of education. Professor Sarah-Jayne Blakemore of University College, London, has suggested that for teenagers the school day should begin a couple of hours later to take into account that their brains are in the grip of hormonal changes that impact on their diurnal and nocturnal rhythm. Melatonin is naturally produced by our brains in the evenings, to help us feel ready for bed. In the teenage brain it is produced a couple of hours later than it is in early childhood or later adulthood. Teenagers are night owls (at the risk of generalising), so forcing them to get up at seven a.m. to go to school, when their brains say they should still be asleep, makes lack of motivation, lack of attention and poor self-discipline more likely. It's easy to think of reasons why it would be 'impossible' to alter an institution's timetable, but perhaps we should be fostering a flexible school day in the same way that we are embracing more flexible working. Sarah-Jayne also advocates postponing testing at the age of sixteen. The current GCSE exam takes place in the middle of a period of intense change in the structure of the teenage brain. The significant fall in grey matter and a corresponding increase in white matter I mentioned earlier has a massive impact on decision-making, planning and self-awareness. Not all teenagers are adversely affected by the existing arrangements, of course, but for those who are more vulnerable, radical changes of this nature may make it easier for them to achieve highly significant positive life outcomes.

Educational neuroscience is not solely focused on the young. Professor Eleanor Maguire, also of University College, London, studied the brains of London taxi drivers to understand more about how learning in adult life affects brain structure. As is well known, London cabbies have a larger than average hippocampus because they have to acquire 'the Knowledge' to be granted their licence. She wanted to study this phenomenon in more detail. It transpired that only about half of the potential cabbies who spent a year studying for the test passed it. The successful individuals indeed had a greater volume of grey matter in the hippocampus, the area involved in memory and navigation. (Incidentally, in a clear demonstration of the maxim 'use it or lose it', the cabbies' hippocampus shrank to a closer-to-average volume when they retired.)

Eleanor wanted to understand why only half of the cabbies pass the test. Do some individuals have limits to their hippocampal plasticity? Do some have a maximum hippocampus volume that is below the limits of acquiring the Knowledge? Teasing apart an individual brain's inherent biological constraints on plasticity is a field that many educational neuroscientists are now focusing on. There are no hard answers as yet but it will be interesting to see how the work develops over the next few years.

In the meantime, the more neuroscience confirms about the benefits of education, the more lifelong learning is touted as an elixir for prolonged brain health. A number of robust studies have demonstrated that ongoing intellectual activity helps ward off dementia and results in lower doses of medication being dispensed to elderly patients in care homes.

Thinking beyond the practicalities of school days or the proven benefits of a Sudoku habit in later life, we arrive at the as-yet-untested realm of incorporating brain training into mainstream education and lifelong learning programmes to foster flexibility of thought, creativity and problem-solving. This is the kind of thing we saw Jonas Kaplan working on as he experimented with techniques to help people regulate their emotions in order to change their minds. Is there an argument for saying schools could or should seek to equip their pupils with proficiency in these techniques? We hear constantly about how, in the age of automation, the skills we will need are emotional intelligence, resilience, creativity and problem-solving but traditional education is still catching up with what that means in practice. Perhaps we should be looking to the experimental fringes of neuroscience for new ideas on how to educate our brains for the future?

One area of public life where neuroscientific insight into the limits of our capacity to act as free agents has already caused speculation is the legal system. How does neuroscience challenge the concept of personal responsibility in a legal sense? In one readily imaginable and anxiety-inducing scenario, neuroscience is appropriated by offenders and their counsels as a defence. The notions of personal agency and individual responsibility on which the entire system of justice depend are eroded by people blaming their neurobiology for their behaviour, claiming it made them act in a particular way. Pushed to its logical endpoint, this scenario sees violent criminals getting away with it because they were able to evade responsibility for their actions by insisting their brain made them do it.

It should go without saying that neither I nor any other neuroscientist I know has ever suggested for a moment that dangerous people should be allowed to escape justice because of their neurobiology. Thankfully there is no prospect of that happening. But neuroscience can provide us with vital nuances to refine our conception of morality, guilt and punishment, and we as a society need to explore them.

To begin with the most fundamental issue of all, the age for criminal responsibility within the English judicial system is ten. From that age, and for the most serious crimes, such as rape and murder, a child is accountable for their behaviour in the same way as an adult, and is deemed sufficiently mature to stand trial in a Crown Court. As we discovered earlier, the standard pattern of brain development sees vast swathes of change and increasing connectivity between the neural circuits involved in impulsivity, decision-making and emotional response occurring well into somebody's early twenties. A ten-year-old brain hasn't even begun on the massive shifts that occur during adolescence. Does it really make sense to say a person of this age has the same capacity for reflection or judgement as a forty-year-old and to treat them in the same way?

I am not suggesting that juveniles who have committed murder should not be locked up. Public safety must be the top priority. Perhaps, though, our instinct to punish, though eminently under-standable in one context, obscures an understanding that a child's brain is qualitatively different from an adult's. Lesser crimes are tried in the juvenile courts, where a sliding scale is used in sentencing, depending on the age of the offender at the time of the crime. It seems that the law in fact recognises that a child is not a mini-adult, except where the crime is so emotive, so heinous, that it erodes our capacity to remember that upholding public safety is not the same as exacting punishment or even enacting justice.

The age-based threshold of criminal responsibility for federal crimes in the United States is eleven, but the country also has an intelligence-based threshold for application of the death penalty. If convicted of a crime for which the death penalty is applicable, an individual must sit a battery of intelligence tests. If the person scores above 70 for any domain of intelligence tested (for reference, the mean population IQ score is 100, 'genius' marked at above 140), they will be executed unless their counsel can find other grounds for appeal. More than 20 per cent of those currently waiting on Death Row would not be executed if the law were based on their average score across all the tests, rather than the current situation in which a single incidence of a score higher than 70 means a tick in the box marked 'personal responsibility', making them eligible for the death sentence.

One final point about age thresholds for criminal responsibility is that, of those countries (a minority) where it's set at fifteen or higher, one or two are those we may expect – socially liberal Norway, for example. The rest are countries in the less developed world that have reformed their penal codes in the last ten or so years, during a period when neuroscientific evidence on neural development during childhood and adolescence was available.

Timor-Leste, Mozambique, Brazil and Argentina set sixteen as the age of criminal responsibility. It's eighteen in Colombia. Many of these countries do, admittedly, have much higher murder rates than we do in Western Europe but have nonetheless managed to practise a collective form of emotional reappraisal as they evaluate the new neuroscientific evidence and change their minds about their justice system.

Beyond the question of thresholds for criminal responsibility, how else can neuroscience contribute to our definition of the concept of agency, sometimes in the most challenging and distressing circumstances? Recognised doctrines within the legal system provide exceptions (the defence of 'diminished responsibility', for example) but otherwise the starting point of the law is that we are considered wholly responsible for our actions. As we learn more about how our behaviours are forged, should this really be the default position?

In order to explore this, it might be helpful to consider a recent landmark case where neuroscience was taken into account during sentencing. An American in his forties developed unusual sexual-arousal behaviours and started making advances on his stepdaughter and collecting child pornography. He was diagnosed with paedophilia, removed from his family home and convicted of child molestation. He was first ordered to attend a rehabilitation programme for sex offenders or face a jail term, but he was expelled from the programme for being unable to stop himself soliciting sexual favours from staff and other participants. The evening before he was due to be sentenced, he was admitted to hospital with a headache and balance problems. An MRI scan revealed a cancerous tumour that had displaced the right orbitofrontal cortex of his brain, an area involved in the regulation of social behaviour. If somebody sustains damage to this area in early life, their ability to make moral judgements is impaired and an antisocial personality may result. The risks are lessened if the damage occurs later in life, but decision-making and impulse control are routinely impacted, and sometimes socio-pathic behaviour results.

The tumour was removed and sentencing deferred. After a couple of days the man's balance returned to normal. A week later he began and subsequently completed a Sexaholics Anonymous programme. Seven months later he was declared no longer a threat to his stepdaughter and allowed to return home. But a year after that, the man reported that he again had persistent headaches and was again collecting child pornography. The tumour was back. A second surgery was performed and the patient's behaviour again returned to normal.

Now, I'm not suggesting that all cases are as open-and-shut as this one was. As far as biological determinism for behaviour goes, it doesn't get much clearer than a brain tumour in a region involved in social and moral behaviour resulting in dramatic changes in urges and action. Remove the tumour, remove the crime. It may be tempting to speculate that there is a neurobio-logical component to all criminal behaviour. After all, the possibility that we may one day be able to excise criminality from our society is alluring. But unfortunately this hope, this confidence in biology and determinism, is possibly unrealistic.

Numerous genetic studies have tried to reveal which genes are linked to criminality. One vast study yielded huge amounts of data from its investigation of more than two thousand twins. There does indeed appear to be strong evidence for heritability in antisocial behaviour, but the story is so complicated, so inter-twined with environment, culture and life experiences, that it is not yet possible to use this information to predict who will commit a criminal act. In fact, although trial judges and parole boards in the US are already using computer algorithms to try to predict who is likely to reoffend, genetic-screening data is currently not included in the database. Neither is it admissible as evidence for consideration by the courts.

That is, with one exception. A single DNA allele is heavily implicated in violent behaviour. It's the metabolising enzyme monoamine oxidase A, or MAOA. Its function is to degrade neurotransmitters, including our old friend dopamine, plus norepinephrine and serotonin. It has been repeatedly shown that low levels of MAOA are associated with aggressiveness in young boys raised in abusive environments.

We still don't know exactly why this is but model organisms – mice in one case – genetically engineered to have low levels of MAOA seem to exhibit striking aggressive behaviours. During a mouse's first days of life, brain scans revealed that their pleasure, happiness and motivation brain chemicals were a whopping ten times greater than expected, potentially desensitising the receptors to these transmitters and dramatically affecting wiring of the circuits involved in emotions and emotional regulation.

Back to humans, and when you scan the adult brains of low-MAOA individuals you can see how their brain activity and region volume are different from those of the majority of the population. When these people view ambivalent images that may or may not suggest threat, they perceive social rejection and insulting behaviour, and their brain profiles show decreased inhibitory control. Effectively, they are less able to assess a situation as non-hostile and choose not to react.

So robust is this association between MAOA and antisocial behaviour that it has been used as evidence during sentencing in the United States. In 2009 an individual who had been convicted of first-degree murder had his sentence reduced from the death penalty to thirty-two years in prison. The man's lawyer demonstrated that his client had the genetic variation for extremely low levels of MAOA and had been abused as a child. The combination of biological and environmental factors was deemed to be enough of a mitigating circumstance to substantially alter his level of personal responsibility for his actions.

Genetic screening is getting cheaper, faster and easier. Neuroscience is demonstrating that adults who have suffered from adverse early-life experiences are more likely to exhibit elevated levels of risk-taking behaviours, akin to the behavioural characteristics of adolescence. They are also less likely to be deterred by the current criminal-justice system for punishment. What, then, should we as a society do with this information? As we discover more, will we be able to develop intervention programmes that target those young people who are vulnerable to criminality? Or work with prisoners to support them to make changes to their behaviour that take into account their own genetics, cognitive habits and life circumstances, with the aim of reducing reoffending? The more we learn from neuroscience, the more it seems that the aim of deterring criminals may be better served by shifting away from a legal model whose primary aim is to punish towards one that treats the problem of criminality as it does public health. Just as obesity is socially contagious, perhaps criminality is, too? If so, then perhaps the prison system is liable to promote rather than protect against its spread.

Treating violence as a public-health issue rather than a legal one is starting to happen in some cities around the world. In the UK, Strathclyde Police, with the support of the Scottish government, have pioneered this approach. In 2005 the police high command decided to do something different in their attempt to reduce violent crime in Glasgow. They looked to Chicago, where epidemiologist Gary Slutkin had applied his insights into the spread of TB and cholera through Somalian refugee camps to Chicago's homicide epidemic. Slutkin recruited what he termed 'reliable messengers'

– convicted but reformed criminals from the neighbourhood he was targeting – to get out on the ground and behave as role models to the vulnerable populations, intervening if necessary and offering practical assistance with everything from accessing treatment for addiction to transport to job interviews. Everywhere he launched a project, the murder rate dropped by at least 40 per cent within a year. The result in Glasgow was the Violence Reduction Unit, which aims to treat violence as something to be cured rather than punished. Glasgow's murder rate has dropped by 60 per cent since the unit's formation in 2005. It's not possible to attribute that change solely to the unit, but the statistics are nonetheless impressive. At some point, perhaps it will be possible to complement the VRU's epidemiology model and its insights drawn from behavioural psychology with genome testing for individual programme participants, or brain-training techniques.

The emerging neuroscience of compassion

For us to decide to pursue such an approach more broadly across society, we would need to foster in ourselves and others a compassionate and collaborative, curious and non-judgemental mindset. We would need to be able to say, 'There is robust scientific evidence that such a mindset is not exceptional, not the invention of a liberal or pacifist imagination. It exists widely throughout the population and, like any trait, it can be encouraged rather than stifled.' This, I suspect, is precisely the sort of thing of which Rowan Williams, my mentor in the pursuit of a more collaborative and open-minded approach to life in general, would approve.

There is, of course, plenty of science to say that Rowan is absolutely right to suggest that linguistic exchange and cultural production support collective cohesion and behaviour change. In fact, one theory has it that language evolved partly as a way round the numerous glitches in our individual brain's processing power. Communicating our personal realities, sharing our experiences and ideas, produces a more reliable working model of how the world functions, and fosters group-level advancement.

The eminent ethologist Richard Dawkins argued in his groundbreaking 1976 book, *The Selfish Gene*, for the existence of a 'unit of cultural transmission' that he termed a 'meme'. The meme need not be language-based, though it often is, but it is always socially spread. Think of it as novel behaviour, passed among peers and down generations. Significant memes might include everything from the skill of fire-lighting to the concept of gender equality. Much as one can inherit genes from each parent, it is suggested that individuals acquire memes through imitating what they observe around them. Those memes replicate, compete and mutate. A meme is deemed successful if it persists across generations and proves adaptable enough to be adopted by a large number of hosts.

There has been criticism of 'memetics' from social scientists and philosophers, who point out that memes, unlike genes, do not have a definite coded identity. They are more nebulous than that so their spread is more chaotic than the trackable evolution of genes, with their absolutely specific DNA coding and clearly observable mechanisms for mutation and selection. Ideas and cultural practices certainly do spread through a society and across generations but, as the late German-American evolutionary biologist Ernst Mayr put it, is there really any need to use the word 'meme' when 'concept' had already served generations of social geographers and cultural historians perfectly well as they set about analysing the spread of ideas across space and time?

What's certainly true is that, over millennia, human beings

have created a vast plethora of activities that cultivate and accelerate the dispersal of concepts (or memes, ideas, behaviours). Conversing with your neighbours during an evening stroll through town, say, or storytelling round a camp fire, practising the visual arts or playing music, visiting nightclubs or bars: every social gathering and artistic expression allows interactions between individuals that tap into our ability to imagine scenarios outside our direct repertoire of experience and to soak in a new way of looking at the world. Neuro-imaging techniques demonstrate that there are dramatic and sustained changes in the brain as exposure to these methods for 'meme contagion' increases. The more a person participates in such activities, the more connectivity in their brain.

It should be noted that the ability to transmit novel ideas and individual perspectives is not limited to humans. Other species also forge communities that foster adaptations in behaviour. In the 1940s scientists observed that when one chimp washed potatoes in a stream before eating them, others followed the example and before long potato-washing was the new social norm. City-dwelling crows have learned to drop nuts onto the road by a pedestrian crossing and wait for a passing car to break them open. The birds then press the crossing button with their beak to bring the traffic to a halt so that they can retrieve the now accessible nut. This behaviour has been observed at numerous sites across different cities. Then there's the humble honey bee and its waggle dance, which we encountered earlier (along with the not-so-humble cocaine-fuelled version). The bee employs its incredible figure-of-eight waggle dance to communicate with other members of the colony. Information about the direction of and distance to untapped sources of pollen is transmitted through movement and gesture. Even trees and plants communicate with each other by hacking into the fungal networks underfoot that have been termed, somewhat whimsically, the 'wood-wide web'. Like our internet, it has its dark side. Some orchids tap into the system to steal resources from nearby trees, while black walnuts spread toxic chemicals through the network to sabotage neighbours and claim more nutrients and sunshine for themselves.

But though communication among animals and even plants is standard, and behaviour of all kinds is socially contagious within animal groups, humans are consummate generators of memes. We have the tools of infinitely sophisticated language and exponentially evolving technologies with which to communicate and transmit ideas. Social media in particular and web-based information-sharing in general is a staggeringly powerful mechanism for the spread of memes. The implications for our sense of personal agency and wider social functioning are still barely understood, as we saw when we found that altering the content of a person's Facebook feed alters their emotional state. Cognitive functioning, decision-making and even choice of political candidate are all potentially up for grabs if our mood can be manipulated by a powerful lobby.

Which brings me back to the idea that neuroscience is, like any other body of biological knowledge, value-neutral. Its application is what determines its effect in the real world. And if we dread the idea that our brains may be manipulated at a distance for political ends, making us less than fully human in the process, we must engage with the possibility of reclaiming our sense of agency via neuroscience's tools. Communication and collaboration are defining human traits and the technology exists to embed such behaviours in our societies as never before. Compassion is, likewise, at least as innate as self-interest. But, of course, we must decide we want to foster a mindset that supports such collectivist values – which means we need a neuroscience of compassion and cooperation. We need to know more about the neurobiological basis of altruism.

The term 'altruism' was coined by the French philosopher Auguste Comte in the 1850s when he developed his doctrine that individuals had an ethical obligation to place the needs of others over their self-interest. However, the branch of philosophy known as ethics is far older than Comte. There has always been debate in philosophical, theological, political and latterly biological circles as to whether humans are truly capable of 'living for others', as Comte put it, or whether even the impulse to kindness is at bottom a selfish one. David Hume and Jean-Jacques Rousseau contended that humankind is unselfish, but Thomas Hobbes asserted that humans possess a natural 'universal selfishness' that drives every aspect of behaviour. Imagine, for example, you offer to help your friend move house. Is this really an altruistic act? Or are you offering in the hope that others in the friendship circle will notice your apparent altruism and your social credit will soar? Are you engaging in what cognitive scientists call 'future planning', banking on your offer of assistance resulting in offers of help if you need it in the future?

Richard Dawkins took the meme of innate human selfishness and refined it in *The Selfish Gene*. 'We are survival machines,' he asserted, 'robot vehicles blindly programmed to preserve the selfish molecules known as our genes.' That bullish view of innate and overriding self-interest has been widely challenged since the book's publication. Dawkins himself wrote in a new foreword to the thirtieth-anniversary edition that he could 'readily see that [the book's title] might give an inadequate impression of its contents'. He said in 2006 that, with hindsight, he should have followed his editor's suggestion and called the book *The Immortal Gene*.

The book was rightly influential, but after a long period of critique of selfishness as an innate quality, it has been superseded in more recent times by the study of an almost opposing behaviour: compassion. For example, in 2005 the Dalai Lama, whose primary teachings include the premise that compassion for others lies at the heart of our individual self-love and happiness, was invited to give the keynote speech at the annual conference of the Society for Neuroscience. Here, more than 31,000 neuroscientists from around the world converged. 'Modern neuroscience has developed a rich understanding of the brain mechanisms that are associated with both attention and emotion,' he said. The Eastern contemplative tradition, given its long history of interest in the practice of mental training, 'offers on the other hand practical techniques for refining attention, and regulating and transforming emotion. The meeting of modern neuroscience and Buddhist contemplative discipline, therefore, could lead to further study of the impact of intentional mental activity on the brain circuits that have been identified as critical for specific mental processes.' Three years later, at Stanford University School of Medicine, a Centre for Compassion and Altruism Research and Education was founded, with the explicit goal of conducting rigorous scientific studies of compassion and altruistic behaviour.

So, what can science now tell us about altruism and compassion? To what extent does it have a biological basis? Are there specific genes or brain regions implicated in this behaviour? Are some individuals born predisposed to be inherently selfless and others pathologically selfish? And is there anything that can help shift the population as a whole towards a more compassionate mindset that is more likely to place value on the emotions of others? Although these questions have been asked in various forms for hundreds of years, neuroscience has only attempted to contribute to answering them over the last fifteen or so and studies are in their infancy.

That said, one fascinating piece of research suggests that individuals lie on a spectrum from selfless to selfish, with those who enact zealous altruism at one end while the clinically diagnosable psychopaths reside at the other. If you imagine a graph, this spectrum of behaviours makes up the X axis. At one end of the axis are psychopaths who engage in extreme antisocial behaviour, violating the rights of others without remorse. Unsuccessful psychopaths are by definition criminals, but psychopathic traits have also been associated with high-achieving individuals in politics, medicine and business. Bold, hard-nosed actions can be instrumental leadership skills, while immunity to stress and fear can yield positives for a society, which perhaps helps to explain why these traits have successfully been propagated throughout generations of our species. Genetic studies of more than five thousand sets of twins estimate that heritability for the core behavioural features of psychopathy – callousness and unemotional traits – lies between 40 and 70 per cent, though, as always, we should keep in mind that high heritability does not mean a simple causal link between certain genes and behaviour.

At the opposite end of the spectrum are extreme altruists, those who consistently place the needs of others above their own with no apparent personal benefit. Studies on highly altruistic individuals have suggested a different brain profile. The altruistic behaviour starts with empathy – the ability to imagine what it would be like to be in somebody else's situation, and to share their feelings. Empathy is highly prized in our culture but, as we've already seen, repeated exposure to others' distress without cathartic action can lead to intolerable personal pain, which cramps a person's ability to alleviate another's suffering. I think of compassion as the pragmatic version of empathy – it is not a matter of simply taking on others' emotion: it's also made up of a strong desire to do something practical to help. Compassion is likely to lead to altruistic action. Helping others in distress can calm painful empathetic feelings, which implies that seeing empathy through to its natural conclusion of an altruistic act is the most beneficial course of action for both the person experiencing the empathy, and the person suffering. An altruistic act might contain some element of self-interest, whether we are conscious of it or not.

Given that compassion and altruism are such highly complex behaviours, it is no surprise that a broad network of interacting genes and neural circuits are believed to be involved in laying the foundations for these pro-social behaviours. Scientists have identified genetic variants that determine dopamine and oxytocin levels and are associated with selfless and selfish behaviours respectively. Various brain circuits are implicated, including our old friend the reward pathway, the amygdala (involved in fear response) and the prefrontal cortex, involved in decision-making.

Returning to the graph, the authors suggest that as a population we occupy an inverted U-shaped distribution, where the Y-axis records the percentage of people whose behaviours range from altruistic to psychopathically selfish. The vast majority of any group fall in the broad middle area of the spectrum but it does seem possible to shift the peak of the curve for any given population, depending on social and cultural factors. This means there is potential to cultivate greater levels of compassion within individuals. On a much larger scale, it may be possible, at least in theory, to instigate a revolution in altruism across an entire population. One of the central theses of this book is that it is beneficial to accept our own and each other's ingrained idiosyncrasies and to value our individual perspectives and inherent glitches in processing, while at the same time debating our differing realities. In this way we will collectively get closer to a more nuanced and robust set of beliefs that better serves our needs.

Why we need a compassion-based mindset more than ever

There has probably never been an epoch when people didn't worry about what the future would bring. It can't have been easy to maintain an optimistic outlook during the Black Death or the First World War. The threat of nuclear annihilation during the Cold War years gave my parents nightmares. Today's global challenges range from the biggest refugee crisis since the end of the Second World War to the existential threat of catastrophic climate change. It can help to ease our sense of panic to remember that the world has always had problems, but it's also true that we face a challenging future and that a great deal more individual and collective action will be necessary to generate solutions.

Having just argued that neuroscience is revealing more and more limitations to personal autonomy and free will, I want to address the question of where this leaves us. We've seen that people tend, in general, towards more selfish behaviour when they believe that their free will is limited or non-existent. That said, the argument that humans are innately selfish is increasingly being challenged by the suggestion that we can just as plausibly claim that humanity is innately altruistic. As I've suggested, sweeping generalisations about human nature flatten out individual variation and belie the complexity of how behaviour is generated. Just as an individual's outcomes are shaped by a matrix of innate and experiential factors, so too are our collective outcomes. Belief systems do shift and evolve, under pressure from collective re-evaluation. Groups can and do change their minds.

There is an opportunity, as we move away from formerly dominant ideas of innate selfishness and the power of the autonomous individual, to rethink our sense of what drives us and shapes our outcomes. I find it troubling to contemplate a society of former devotees of free will driven by disillusion to become nihilists and ideologues. That's why there is value, to my mind, in constructing a neuroscientific argument for an innate collective consciousness, and for humanity's potential for altruism and compassion. If we can integrate this concept into our thinking we may be able to shift in the direction of collective action to tackle global issues or, more simply, to listen to our neighbours' opinions.

But how on earth do we go about galvanising a shift towards greater open-mindedness, towards enhanced compassion and altruism? In late 2017 the fantastic *Oxford Handbook of Compassion Science* was published, reviewing findings from this emerging field. I'm going to attempt to distil some of the main conclusions into five tangible tips that we can all, hopefully, use to integrate compassion and improved communication into our everyday lives.

1. **Learn to recognise and talk about your own emotions**
   The act of learning to recognise your own emotions and communicating them positively to others physically changes how you perceive the emotion. Calmly saying, for example, 'I feel angry,' dampens the primitive emotional anger response in the brain and redirects activity to higher cognitive circuits, thereby helping to alleviate the emotional distress of the anger. With a bit of luck the person you express this emotion to will respond compassionately. But even if they do not, simply verbalising the emotion has made space in your brain for you to regain control and act positively and compassionately within yourself. Similarly, some people have found it helpful to be trained in picking up others' emotions without explicit verbal cues, by observing their gestures, expressions and actions. These skills can help cultivate friendships and seed a more

compassionate outlook. So, maybe sit down with friends and practise pulling and reading emotional faces. If nothing else, it could lead to an entertaining evening.

2. Practise compassion meditation

This involves making time for self-reflection with the emphasis on why you like yourself. The idea is to show yourself compassion, despite intimately knowing your flaws. You then turn your attention to the people you love, including them in your compassion and gratitude. Finally, think of the difficult people in your life, those towards whom you feel active hostility. It takes practice but the aim is to wish that they, too, might be filled with loving-kindness and peace. Compassion meditation has been found to increase self-reported mindfulness, happiness, feelings of compassion to self and others and decreased self-reported worry. It can help us to deal with potentially negative or distressing events by bolstering our ability to frame them in a less overwhelming way. It also alters our perspective on what constitutes happiness, shifting the emphasis away from the see-saw activity of our old friend the hedonic reward pathway, interspersed with periods of fear, towards living in a more grounded state of mind.

3. Acknowledge compassion in others

Witnessing other people being altruistic not only improves your feelings of optimism about humanity but also boosts your own desire to help others. It fosters moral elevation, a powerful emotion that is part of the awe family. This emotion increases the prefrontal cortex's control over more primitive emotional circuits responsible for the fight-or-flight response, so that you are able to deploy more executive decision-making. Moral elevation also raises oxytocin levels, lowers cortisol levels and increases neural plasticity, facilitating the integration of unexpected experiences into a person's understanding of the world. Collectively, this positive emotion of moral elevation helps to promote a 'pay it forward' mentality. Taking the time to notice acts of kindness can help compassion spread across society.

4. Practise gratitude

We live in highly individualistic societies but we have not evolved to be self-sufficient. Our pro-social brains have developed in line with the benefits that derive from our interdependence on each other. Essentially, having supportive relationships in place helps us to survive. The simple act of practising gratitude and appreciation of others helps us to value that support. Before going to bed each night I mentally list the three things from the day that I am grateful for. This helps me to remember to express my gratitude to the individual responsible, ends the day on a positive note and gets me thinking of ways to induce this feeling in others by acting with kindness myself.

5. Be a compassion-focused parent

Raising children in a context that strengthens compassion can help them to develop their own positive support network in later life, thereby helping to spread altruism further afield and across generations. Children learn which emotions are acceptable and how to regulate them by observing their caregivers. Being aware and in control of your own emotions by using techniques, such as slow breathing, to calm anger is beneficial for you and for your children. Similarly, studies have shown that the children of parents who make time to practise their own self-care also benefit over the long term. Eat well, take time out to exercise and see your friends; relax with hobbies or practise meditation.

I find it tempting to speculate that humanity has been driven to develop technologies that break down our geographical boundaries and open us up to new ideas. Our increasingly interconnected world places us in the unprecedented situation of being able to communicate our ideas and needs more easily than ever before. We are also more aware than ever before of the suffering that is occurring around the world, be it in refugee camps, plastic-filled oceans or in the wake of natural disasters. We have developed the tools that increase our exposure to those suffering and we have improved our understanding of why certain people inflict pain on themselves or others.

I like to think that we can use this information positively by setting it in the context of our enhanced understanding of how our behaviours are generated. The neuroscience of behaviour could be used as a powerful force to manipulate us into feeling permanently afraid, which would inevitably result in a more divided society. Or we could choose to confront fear and division. We could employ the advances in understanding of our predisposed behaviour intelligently, to inform how our society generates policies that will positively shape our education, health, criminal justice and communications systems for the future. I hope we choose the latter option.

Epilogue

One day in the spring of 2018, I was cycling across the commons in Cambridge on my way to talk to a neuroscientist about fate, resilience and free will. The parkland was looking particularly idyllic and I was just contemplating some cowslips when in the near distance I noticed a colleague, Dr Mike Anderson, also pedalling on his way to work. We hadn't bumped into each other in more than a year so I called his name, caught him up and we cycled side by side as we exchanged our news. It turned out that he had been busy. Before I mentioned what was keeping me occupied, he beamed as he told me of the arrival of his first child, who was then seven months old.

Once I had congratulated him, he told me a story. His wife is originally from Korea and the previous weekend they had been to the first birthday party of the child of some friends of theirs, a child of the same mixed Korean and British heritage. He excitedly told me what this entailed. In Korea, a child's first birthday is a particularly big deal, and the party is planned months in advance. It features an ancient tradition, key to the celebrations, where the infant is presented with a tray of objects, each signifying a specific life trajectory. The child is encouraged to grab something from the tray and the parents and guests watch with bated breath as he or she selects their fate.

The couple at this party, Mike told me, had been willing their baby to pick up the stethoscope, which represented a career in medicine. And which object did the child select? 'Oh, the stethoscope of course!' Mike replied, with a smile.

'What will you be willing your son to pick up?' I asked him. Mike grinned, chuckling to himself. 'The book indicating a scholar, naturally!'

As we reached the junction in our respective journeys, Mike asked me what I had been up to. I was reeling somewhat from this chance encounter that had provided me with an unexpected parable about the persistent power of the idea of fate. I couldn't quite articulate all of that, though, so I mumbled something vague as we said our goodbyes.

Mike's story charmed me. There was something so disarming about the symbolism of the objects, the proud but anxious parents and the doctor in the making, waving his stethoscope. Human beings love to tell each other stories, about themselves and each other, about how life might turn out and whether any of it is under our control. Fate is out of date, yet it recurs in our stories, our modes of speech, and on a tray of symbolic objects at a child's first birthday party.

My work on this book has confirmed my conviction that fate is still meaningful, and that it's located at every branching junction in the vast connectome of our brain, generated as each and every synapse, trillions of connections, spark and fizz with our brain's staggering, awe-inspiring power. The more we know about how this contemporary version of fate operates, the more chance we have of working with it, rather than against it.

Neuroscience suggests that the answer to the question of how much we can control our lives is incredibly complicated and nuanced, but essentially the more we learn about our brains the stronger the argument for predetermined fate. We are beginning to understand how vast swathes of our complex behaviours are ingrained, handed down the generations by mind-blowing mechanisms, written into our DNA code and through the volume dial for our genes, to direct the construction of the circuitry that makes up our mind. Through inherent processing constraints, our perception of the world and our sense of reality amplify these biological predispositions and confirm us on the life trajectory we were born to. On the flip side, the exceptional plasticity, dynamism and flexibility that also characterise our brain provide scope to alter our behaviour and, potentially, our course. But to break down individual habits requires persistence, as well as self-reflection and the ability to communicate with, and hold compassion for, others.

It is by engaging constructively with difference wherever we find it that our ideas will adapt, as they must for us to thrive. I would like to think it is empowering to discover our own fate, which seems to me to be another way of saying our flaws, our inherent biases and propensities. Perhaps, paradoxically, it will help to prevent powerless rumination and instead enable us to live with greater appreciation for the majesty of what our brains enable us to think and do.

Acknowledgements

Firstly, thank you to the brain – that awe-inspiring majestic beast that creates each of our highly individual worlds.

Many people contributed towards this book – all the researchers mentioned within the pages who generously donated their time to discuss their fields, my previous supervisors Professor Trevor Robbins, Dr Olga Krylova, Dr Jeremy Skepper. Dr Melanie Munro and Dr Peter Maycox who inspired me with their knowledge and enthusiasm during the beginning of my career, and the countless neuroscience friends I've made along the way, whose passion for the brain is gloriously infectious.

Sincere thanks also go to Caroline Michel, literary agent extraordinaire, who made this project possible, Peter Florence for creating the Hay Literary Festival and providing a phenomenal platform celebrating the art of writing and reading, the unfailingly positively crafted guidance from Rowena Webb and Maddy Price at Hodder and to writing mentor Helen Coyle, with whom I spent hundreds of hours discussing how best to navigate the nuances of this book. I learnt a lot and miss having an excuse to discuss the brain, and life, with you!

Sincere gratitude goes to Dr Rogier Kievit, Dr Ari Ercole and Nicky Buckley for generously taking the time to read this manuscript, offering insightful comments and guidance, and to Magdalene College, University of Cambridge for providing many stimulating conversations and diversions. And finally thanks to my family, particularly my mum and dad, and my friends for all their support over the years. Many names stand out but particular thanks cheerfully go to Captain Mark Nash who kept my course on an even keel.

Finally, heartfelt acknowledgements go to the children who lived in the psychiatric hospital two decades ago.

References

Works are referred to in the order in which citations appear in the text.

1. Free Will or Fate?

Sapolsky, Robert M. (2017) *Behave: The biology of humans at our best and at our worst*, Penguin Press.

Kahneman, Daniel (2012) (reprint edition) *Thinking, Fast and Slow*, Penguin Press.

Satel, Sally and Lilienfeld, Scott O. (2015) *Brainwashed: The Seductive Appeal of Mindless Neuroscience*, Basic Books.

Royal Society (2011) *Brain Waves Module 1: Neuroscience, Society and Policy*, London.

—— *Brain Waves Module 2: Neuroscience: implications for education and lifelong learning*, London.

—— *Brain Waves Module 3: Neuroscience, conflict and security*, London.

—— *Brain Waves Module 4: Neuroscience and the law*, London.

Hilker, R. *et al.* (2017) 'Heritability of Schizophrenia and Schizophrenia Spectrum Based on the Nationwide Danish Twin Register', *Biological Psychiatry*, 83(6): 492–8

2. The Developing Brain

Sterne, Laurence (1996) (new edition) *Tristram Shandy*, Wordsworth Editions.

Saint-Georges, C. *et al.* (2013) 'Motherese in Interaction: At the Cross-Road of Emotion and Cognition? (A Systematic Review)', *PLoS One*; 8(10):e78103.

Critchlow, Hannah (2018) *Consciousness: A LadyBird Expert Book*, Michael Joseph, Penguin.

Leong, V. *et al.* (2017) 'Speaker gaze increases information coupling between infant and adult brains.', *PNAS*, 114(50):13290–5.

Mischel, W. *et al.* (1989) 'Delay of gratification in children.', *Science*, 244:933–8.

Mischel, W. *et al.* (1972) 'Cognitive and attentional mechanisms in delay of gratification', *Journal of Personality and Social Psychology*, 21(2):204–218.

Watts, T. W. *et al.* (2018) 'Revisiting the Marshmallow Test: A Conceptual Replication Investigating Links Between Early Delay of Gratification and Later Outcomes.', *Psychol Sci.*, 29(7):1159–77.

Caspi, A. *et al.* (2005) 'Personality Development: Stability and Change.', *Annu. Rev. Psychol.*, 56:453–84.

Blakemore, S. J. (2018) *Inventing Ourselves: The Secret Life of the Teenage Brain*, Doubleday, an imprint of Transworld Publishers, Penguin Random House.

Wenger, E. *et al.* (2017) 'Expansion and Renormalization of Human Brain Structure During Skill Acquisition.', *Trends in Cognitive Neuroscience (Opinion)*, 21, 12:930–9.

El-Boustani, S. *et al.* (2018) 'Locally coordinated synaptic plasticity of visual cortex neurons in vivo.', *Science*, 360,6395:1349–54.

Matthews, F. E. *et al.* (2016) 'A two-decade dementia incidence comparison from the Cognitive Function and Ageing Studies I and II.', *Nature Communications*, 7:11398.

Gerstorf, D. *et al.* (2015) 'Secular changes in late-life cognition and well-being: Towards a long bright future with a short brisk ending?', *Psychol Aging.*, 30(2):301–10.

Kempermann, G. *et al.* (1997) 'More hippocampal neurons in adult mice living in an enriched environment.', *Nature*, 386(6624):493–5.

Talan, J. (2018) 'Neurogenesis: Study Sparks Controversy Over Whether Humans Continue to Make New Neurons Throughout Life.', *Neurology Today*, 18,7:62–6.

de Dieuleveult, A.L. *et al.* (2017) 'Effects of Aging in Multisensory Integration: A Systematic Review.', *Front. Aging Neurosci.*, 9:80.

Fuhrmann, D. *et al.* (2018) 'Interactions between mental health and memory during ageing.', Cambridge Neuroscience Seminar poster prize.

Henson, R. N. A. *et al.* (2016) 'Multiple determinants of ageing memories.', *Scientific Reports*, 6:32527.

3. The Hungry Brain

Livet, J. *et al.* (2007) 'Transgenic strategies for combinatorial expression of fluorescent proteins in the nervous system.', *Nature*, 450 (7166):56–62.

Cording, A. C. (2017) 'Targeted kinase inhibition relieves slowness and tremor in a *Drosophila* model of LRRK2 Parkinson's disease.', *NPJ Parkinson's Disease*, 3:34.

Robbins, T., Everitt, B., Nutt, D. (eds.), (2010) *The Neurobiology of Addiction* (Philosophical Transactions of the Royal Society of London. Series B, Biological Sciences), OUP, Oxford.

Gulati, P. *et al.* (2013) 'Role for the obesity-related FTO gene in the cellular sensing of amino acids.', *Proc Natl Acad Sci USA.*, 110(7):2557–62.

Gulati, P. *et al.* (2013) 'The biology of FTO: from nucleic acid demethylase to amino acid sensor.' *Diabetologia*, 56(10):2113–21.

Loos, R.J. *et al.* (2014) 'The bigger picture of FTO: the first GWAS-identified obesity gene.', *Nat Rev Endocrinol.*, 10(1):51–61.

Hetherington, M.M. (2017) 'Understanding infant eating behaviour: Lessons learned from observation.', *Physiology & Behavior*, 176:117–24.

—— (2016), 'Nutrition in the early years – laying the foundations of healthy eating.', *Nutrition Bulletin* (editorial), 41:310–13.

Chambers L. *et al*. (2016) 'Reaching consensus on a "vegetables first" approach to complementary feeding.', *British Nutrition Bulletin*, 41:270–6.

Nekitsing C. *et al*. (2018) 'Developing healthy food preferences in preschool children through taste exposure, sensory learning and nutrition education.', *Current Obesity Reports*, 7:60–7.

Kleinman, R.E. *et al*. (2017) 'The Role of Innate Sweet Taste Perception in Supporting a Nutrient-dense Diet for Toddlers, 12 to 24 Months: Roundtable proceedings.', *Nutrition Today*, 52:S14–24.

Hetherington, M.M. *et al*. (2015) 'A step-by-step introduction to vegetables at the beginning of complementary feeding: the effects of early and repeated exposure.', *Appetite*, 84:280–90.

Eat Right Now: https://goeatrightnow.com

Schulz, L. C. (2010) 'The Dutch Hunger Winter and the developmental origins of health and disease.', *PNAS*, 107 (39):16757–8.

Tobi, E. W. *et al*. (2014) 'DNA methylation signatures link prenatal famine exposure to growth and metabolism.', *Nature Communications*, 5:5592.

Dias, B.G. *et al*. (2014) 'Parental olfactory experience influences behavior and neural structure in subsequent generations', *Nature Neuroscience*, 17(1):89–96.

Keifer, Jr, O.P. *et al*. (2015) 'Voxel-based morphometry predicts shifts in dendritic spine density and morphology with auditory fear conditioning.', *Nature Communications*, 6:7582.

Boyden, E. S. *et al*. (2005) 'Millisecond-timescale, genetically targeted optical control of neural activity.', *Nat. Neurosci*. 8 (9):1263–8.

Deisseroth, K. *et al*. (2006) 'Next-Generation Optical Technologies for Illuminating Genetically Targeted Brain Circuits', *Journal of Neuroscience*, 26 (41):10380–6.

Karnani, M.M. *et al*. (2011) 'Activation of central orexin/hypocretin neurons by dietary amino acids', *Neuron*, 72 (4):616–29.

Benabid, A. L. (2003) 'Deep brain stimulation for Parkinson's disease.', *Current Opinion in Neurobiology*, 13, 6:696–706.

Hollands G.J. *et al*. (2016) 'The impact of communicating genetic risks of disease on risk-reducing health behaviour: systematic review with meta-analysis.', *BMJ*; 352:i1102.

Marteau, T.M. (2018) 'Changing minds about changing behaviour.', *Lancet*, 391:116–17.

4. The Caring Brain

Miller, G. *et al.* (2007) 'Ovulatory cycle effects on tip earnings by lap dancers: economic evidence for human estrus?', *Evolution & Human Behavior*, 28, 6:375–81. https://doi.org/10.1016/j.evolhumbehav.2007.06.002

Wedekind, C. *et al.* (1995) 'MHC-dependent mate preferences in humans.', *Proc Biol Sci.*, 260 (1359):245–9.

Ober, C. *et al.* (2017) 'Immune development and environment: lessons from Amish and Hutterite children.', *Curr Opin Immunol.*, 48:51–60.

Kohl, J. *et al.* (2013) 'A Bidirectional Circuit Switch Reroutes Pheromone Signals in Male and Female Brains.', *Cell*, 155–7:1610–23.

Grosjean, Y. *et al.* (2011) 'An olfactory receptor for food-derived odours promotes male courtship in Drosophila.', *Nature*, 478:236–40.

Cachero, S. *et al.* (2010) 'Sexual dimorphism in the fly brain.', *Current Biology*, 20(18) 1589–1601.

Bogaert, A. F. *et al.* (2017) 'Male homosexuality and maternal immune responsivity to the Y-linked protein NLGN4Y.', *PNAS*, 115(2):302–6.

Yule, M.A. (2014) 'Biological markers of asexuality: Handedness, birth order, and finger length ratios in self-identified asexual men and women.', *Arch Sex Behav.*, 43(2):299–310.

Kohl, J. *et al* (2018) 'Neural control of parental behaviors.', *Curr Opin Neurobiol.*, 49:116–22.

Kohl, J. *et al.* (2018) 'Functional circuit architecture underlying parental behaviour.', *Nature*, 556(7701):326–31.

Kohl, J. (2017) 'The neurobiology of parenting: A neural circuit perspective.', *Bioessays*, 39(1):1–11.

Fine, Cordelia (2011) *Delusions of Gender: The Real Science Behind Sex Differences*, Icon Books.

—— (2018) *Testosterone Rex: Unmaking the Myths of Our Gendered Minds*, Icon Books.

Dunbar, Robin (2012) *The Science of Love*, John Wiley & Sons, Faber.

Holt-Lunstad, J. *et al.* (2010) 'Social Relationships and Mortality Risk: A Meta-analytic Review.', *PLoS Med.*, 7(7):e1000316.

Dunbar, R.I.M. (2018) 'The Anatomy of Friendship.', *Trends Cogn Sci.*, 22(1):32–51.

Pearce, E. *et al.* (2017) 'Variation in the ß-endorphin, oxytocin, and dopamine receptor genes is associated with different dimensions of human sociality.', *Proc Natl Acad Sci USA*, 114(20):5300–5.

Dahmardeh, M. *et al.* (2017) 'What Shall We Talk about in Farsi?: Content of Everyday Conversations in Iran.', *Hum Nat.*, 28(4):423–33.

Dunbar, R.I.M. (2018) 'The Anatomy of Friendship.', *Trends Cogn Sci.* 22(1):32–51.

Eisenberger, N.I. *et al.* (2006) 'An experimental study of shared sensitivity to physical pain and social rejection.', *Pain*, 126:132–8.

Eisenberger, N.I. *et al.* (2004) 'Why rejection hurts: A common neural alarm system for physical and social pain.', *Trends in Cognitive Sciences*, 8:294–300.

Eisenberger, N.I. *et al.* (2003) 'Does rejection hurt? An fMRI study of social exclusion.', *Science*, 302:290–2.

Shpigler, H.Y. *et al.* (2017) 'Deep evolutionary conservation of autism-related genes.', *Proc Natl Acad Sci USA*, 114(36):9653–8.

Robinson, G.E. *et al.* (2017) 'Epigenetics and the evolution of instincts.', *Science.* 356(6333):26–7.

Feldman, R. (2017) 'The Neurobiology of Human Attachments.', *Trends Cogn Sci.*, 21(2):80–99.

Barron, A.B. *et al.* (2007) 'Octopamine modulates honey bee dance behavior.' *Proceedings of the National Academy of Sciences*, 104:1703–7.

Barron, A.B. *et al.* (2008) 'Effects of cocaine on honeybee dance behaviour.' *Journal of Experimental Biology*, 212:163–8.

Shpigler, H.Y. *et al.* (2017) 'Deep evolutionary conservation of autism-related genes.', *Proceedings of the National Academy of Sciences*, 114 (36):9653–8.

Robinson, G. E. *et al.* (2005) 'Sociogenomics: Social life in molecular terms.', *Nature Reviews Genetics*, 6:257–70.

Young, R. L. *et al.* (2019) 'Conserved transcriptomic profiles underpin monogamy across vertebrates', *PNAS* 116 (4): 133–6.

5. The Perceiving Brain

Critchlow, Hannah (2018) *Consciousness: A LadyBird Expert Book*, Michael Joseph, Penguin.

Gegenfurtner, K.R. *et al.* (2015) 'The many colours of "the dress"', *Curr Biol.*, 25(13):R543–4.

Wallisch, P. (2017) 'Illumination assumptions account for individual differences in the perceptual interpretation of a profoundly ambiguous stimulus in the color domain: "The dress"', *Journal of Vision.* 17 (4):5.

Gregory R. L. (1997) From: *Phil. Trans. R. Soc. Lond.* B 352:1121–8, with the kind permission of the editor.

Gregory, Richard (1970) *The Intelligent Eye*, Weidenfeld and Nicolson.

Króliczak G. *et al.* (2006) 'Dissociation of perception and action unmasked by the hollow-face illusion'. *Brain Res.* 1080 (1):9–16.

Dima, D. *et al.* (2009) 'Understanding why patients with schizophrenia do not perceive the hollow-mask illusion using dynamic causal modelling.', *NeuroImage*, 46(4):1180–6.

Frith, C. D. (2015) *The Cognitive Neuropsychology of Schizophrenia (Classic Edition)*, Psychology Press & Routledge Classic Editions.

Frith, C. D. *et al.* (2018) 'Volition and the Brain – Revisiting a Classic Experimental Study', *Science & Society Series: Seminal Neuroscience Papers 1978–2017*, 41, 7:405–7.

Lennox, B.R. (2017) 'Prevalence and clinical characteristics of serum neuronal cell surface antibodies in first-episode psychosis: a case-control study.', *Lancet Psychiatry*, 4(1):42–8.

Zandi, M.S. *et al.*(2014) 'Immunotherapy for patients with acute psychosis and serum N-Methyl D-Aspartate receptor (NMDAR) antibodies: a description of a treated case series.', *Schizophr Res.*, 160(1–3):193–5.

Carhart-Harris R.L. *et al.* (2016) 'Neural correlates of the LSD experience revealed by multimodal neuroimaging.', *Proc Natl Acad Sci USA.*, 113(17):4853–8.

Bahrami, B. *et al.* (2010) 'Optimally interacting minds.', *Science*, 329(5995):1081–5.

Frith, C.D. *et al.* (2007) 'Social cognition in humans.', *Curr Biol.*, 17(16):R724–32.

Hofer, S.B. *et al.* (2010) 'Dendritic spines: the stuff that memories are made of?' *Curr Biol.* 20(4):R157–9.

Fine, Cordelia (2011) *Delusions of Gender: The Real Science Behind Sex Differences*, Icon Books.

—— (2018) *Testosterone Rex: Unmaking the Myths of Our Gendered Minds*, Icon Books.

6. The Believing Brain

Shermer, Michael (2011) *The Believing Brain: From Ghosts and Gods to Politics and Conspiracies – How We Construct Beliefs and Reinforce Them as Truths*, Times Books.

Critchlow, Hannah (2018) *Consciousness: A LadyBird Expert Book*, Michael Joseph, Penguin.

http://fcmconference.org/img/
CambridgeDeclarationOnConsciousness.pdf

MacKay, Donald. M. (1991) *Behind the Eye*, Basil Blackwell.

Beauregard, M. *et al.* (2006) 'Neural correlates of a mystical experience in Carmelite nuns.', *Neurosci Lett.*, 405(3):186–90.

Smith, T.B. *et al.* (2003) 'Religiousness and depression: evidence for a main effect and the moderating influence of stressful life events.', *Psychol Bull.*, 129(4):614–36.

Jack, A. I. *et al.* (2016) 'Why Do You Believe in God? Relationships between Religious Belief, Analytic Thinking, Mentalizing and Moral Concern.', *PLoS One*, 11(3):e0149989.

Jeeves, Malcolm and Brown, Warren (2009) *Neuroscience, Psychology and Religion: Illusions, Delusions, and Realities about Human Nature*, Templeton Press.

Schreiber, D. *et al.* (2013) 'Red brain, blue brain: evaluative processes differ in Democrats and Republicans.' *PLoS One*, 8(2):e52970.

Kramer, Adam D.I. *et al.* (2014) 'Experimental evidence of massive-scale emotional contagion through social networks.' *PNAS*, III (24) 8788-90.

Kaplan, J.T. *et al.* (2016) 'Neural correlates of maintaining one's political beliefs in the face of counterevidence.', *Scientific Reports*, 6:39589.

Harris, S. *et al.* (2009) 'The neural correlates of religious and non-religious belief.', *PLoS One*, 4(10):e7272.

Patoine, B. (2009) 'Desperately Seeking Sensation: Fear, Reward, and the Human Need for Novelty: Neuroscience Begins to Shine Light on the Neural Basis of Sensation-Seeking.', Briefing Paper, The Dana Foundation.

Costa, V.D. *et al.* (2014) 'Dopamine modulates novelty seeking behavior during decision making.', *Behav. Neurosci.*, 128(5):556–66.

Molas, S. *et al.* (2017) 'A circuit-based mechanism underlying familiarity signaling and the preference for novelty.', *Nat.Neurosci.*, 20(9):1260–8.

Tang, Y.Y. *et al.* (2015) 'The neuroscience of mindfulness meditation.', *Nature Reviews Neuroscience*, 16(4):213–25.

Galante, J. *et al.* 'Effectiveness of providing university students with a mindfulness-based intervention to increase resilience to stress: a pragmatic randomised controlled trial.' *Lancet Public Health*, 2:PE72–E81.

Shors, T. J. *et al.* (2014) 'Mental and Physical (MAP) Training: A Neurogenesis-Inspired Intervention that Enhances Health in Humans.', *Neurobiol. Learn Mem.*, 115:3–9.

Libet, B. *et al.* (1983) 'Time of Conscious Intention to Act in Relation to Onset of Cerebral Activity (Readiness-Potential)'., *Brain*, 106(3):623–42.

Williams, Rowan (2018) *Being Human: Bodies, Minds, Persons*, SPCK Publishing.

7. The Predictable Brain

Nakamura, A. *et al.* (2018) 'High performance plasma amyloid-ß biomarkers for Alzheimer's disease.', *Nature*, 554:249–54.

https://www.genomicsengland.co.uk/the-100000-genomes-project/

Day, F. R. *et al.* (2016) 'Physical and neurobehavioral determinants of reproductive onset and success.', *Nature Genetics*, 48:617–23.

https://www.bbc.co.uk/news/magazine-37500189

http://nuffieldbioethics.org/project/genome-editing-human-reproduction

http://nuffieldbioethics.org/project/non-invasive-prenatal-testing

Feder, A. *et al.* (2009) 'Psychobiology and molecular genetics of resilience.', *Nat Rev Neurosci.*, 10(6):446–57.

Baker, K. *et al.* (2014) 'Chromosomal microarray analysis – a routine clinical genetic test for patients with schizophrenia.', *Lancet Psychiatry*, 1(5):329–31.

Deary, I.J. (2012) 'Intelligence.', *Annual Review of Psychology*; 63(1):453–82.

Plomin R. *et al.* (2014) 'Genetics and intelligence differences: five special findings.', *Molecular Psychiatry*; 20:98.

Hill, W. *et al.* (2018) 'A combined analysis of genetically correlated traits identifies 187 loci and a role for neurogenesis and myelination in intelligence.', *Molecular Psychiatry*: 1.

Hill, W.D. *et al.* (2018) 'Genomic analysis of family data reveals additional genetic effects on intelligence and personality.', *Molecular Psychiatry*.

Hill, W.D. *et al.* (2016) 'Molecular genetic contributions to social deprivation and household income in UK Biobank.', *Current Biology*, 26(22):3083–9.

Hill, W.D. *et al.* (2019) 'What genome-wide association studies reveal about the association between intelligence and mental health.', *Current Opinion in Psychology*; 27:25–30.

Deary, I.J. *et al.* (2018) 'What genome-wide association studies reveal about the association between intelligence and physical health, illness, and mortality.', *Current Opinion in Psychology*, 27:6–12.

Ritchie, S. (2015) *Intelligence: All that matters*, Hodder & Stoughton.

Spearman, C. (1904) '"General Intelligence" objectively determined and measured.', *Am J Psychol.*,15:201–92.

Calvin, C.M. *et al.* (2017) 'Childhood intelligence in relation to major causes of death in 68 year follow-up: prospective population study.', *BMJ*, 357:j2708.

Ioannidis, K. *et al.* (2018) 'The complex neurobiology of resilient functioning after child maltreatment.' https://doi.org/10.31219/osf.io/3vfqb

Askelund, A. D. *et al.* (2018) 'Positive memory specificity reduces adolescent vulnerability to depression.', doi: https://doi.org/10.1101/329409

Fritz, J. F. *et al.* (2018) 'A systematic review of the social, emotional, cognitive and behavioural factors that benefit mental health in young people with a history of childhood adversity.', Shared last authorship. Preprint, *Frontiers in Psychiatry*, special issue on Resilience.

Plomin, Robert (2018) *Blueprint: How DNA Makes Us Who We Are*, Allen Lane, Penguin Random House, London.

Mitchell, Kevin J. (2018) *Innate: How the Wiring of Our Brains Shapes Who We Are*, Princeton University Press, New Jersey.

8. The Compassionate Brain

Cabinet Office and Behavioural Insights Team (2012) *Test, Learn, Adapt: Developing Public Policy with Randomised Controlled Trials*, London.

Datar, A. *et al.*(2018) 'Association of Exposure to Communities with Higher Ratios of Obesity with Increased Body Mass Index and Risk of Overweight and Obesity Among Parents and Children.', *JAMA Pediatrics*.

https://www.telegraph.co.uk/politics/2018/06/01/supermarket-guilt-lanes-two-for-one-junk-food-offers-will-banned/

Kelly Brownell, professor of psychology and neuroscience and dean of the Sanford School of Public Policy at Duke University in the USA.

Cambridge Public Policy SRI (2017) *The Educated Brain Policy Brief: Late Childhood and Adolescence*, Cambridge.

Blakemore, S. J. (2018) *Inventing Ourselves: The Secret Life of the Teenage Brain*, Doubleday, an imprint of Transworld Publishers, Penguin Random House, London.

Maguire, E.A. *et al.* (2006) 'London taxi drivers and bus drivers: a structural MRI and neuropsychological analysis.', *Hippocampus*, 16(12):1091–1101.

Royal Society (2011) *Brain Waves Module 1: Neuroscience, Society and Policy*, London.

—— *Brain Waves Module 2: Neuroscience: implications for education and lifelong learning*, London.

—— *Brain Waves Module 3: Neuroscience, conflict and security*, London.

—— *Brain Waves Module 4: Neuroscience and the law*, London.

Sapolsky, Robert M. (2017) *Behave: The biology of humans at our best and at our worst*, Penguin Press.

https://www.theguardian.com/news/2018/jul/24/violent-crime-cured-rather-than-punished-scottish-violence-reduction-unit

Godar, Sean C. *et al.* (2016) 'The role of monoamine oxidase A in aggression: current translational developments and future challenges.', *Prog Neuropsychopharmacol Biol Psychiatry*, 69:90–100.

Dawkins, Richard (2016) *The Selfish Gene* (4th edition), OUP.

Mayr, E. (1997) 'The objects of selection.' *PNAS*, 94(6):2091–4.

Critchlow, Hannah (2018) *Consciousness: A LadyBird Expert Book*, Michael Joseph, Penguin.

Rhodes, Christopher J. (2017) 'The whispering world of plants: "The Wood Wide Web".', *Science Progress*, 100, 3:331–7(7).

Sonne, J.W.H. *et al.* (2018) 'Psychopathy to Altruism: Neurobiology of the Selfish-Selfless Spectrum.', *Front Psychol.*, 9:575.

Kosinski, M. *et al.* (2013) 'Private traits and attributes are predictable from digital records of human behaviour.', *PNAS*, 110(15):5802–5.

Kramer, A. d. I. *et al.* (2014) 'Experimental evidence of massive-scale emotional contagion through social networks.', *PNAS*, 111 (24):8788–90.

Bartal, I. Ben-Ami *et al.* (2011) 'Empathy and pro-social behavior in rats.' *Science*, 334:1427–30.

Seppälä, Emma M. *et al.* (2017) Th*e Oxford Handbook of Compassion Science*, OUP USA.

Index